Scientific and Technical Writing Today

from Problem to Proposal

William Magrino

Michael Goeller

RUTGERS, THE STATE UNIVERSITY OF NEW JERSEY

Kendall Hunt
p u b l i s h i n g   c o m p a n y
4050 Westmark Drive • P O Box 1840 • Dubuque IA 52004-1840

www.kendallhunt.com
Send all inquiries to:
4050 Westmark Drive
Dubuque, IA 52004-1840

Printed in the United States of America
10 9 8 7 6 5 4 3 2

Contents

The Project Proposal from the Ground Up

Chapter 1

Someone once said, "those who know *how* will always have a job, but those who know *why* will lead the way." When you write a project proposal, you need to answer two critical "why" questions: "why do this?" and "why *this way* as opposed to some *other* way?" A key premise of this text is that you can only answer those "why" questions through research. Without research, you will not have knowledge (or, at least, you will not be able to persuade other people that you have knowledge), and without knowledge you cannot answer "why" in a way that persuades people to follow you.

We live in a society where knowledge is at the heart of the decision-making process. In this "knowledge society," as Peter Drucker has called it, "knowledge workers" need many complex skills and abilities to get things done. They need to be able to:

- guide their own learning to master new knowledge and skills,

- analyze new situations, assess information needs, and locate that information,

- understand and digest both factual and theoretical material,

- think creatively to combine or improve available ideas,

- harness knowledge to justify a plan,

- develop and explain complex plans of action to others, and

- manage people and resources by putting information into action.

Knowledge workers need to be prepared for the creative challenge of solving problems through research, and they need practice in communicating their research to others. Having the experience of writing a project proposal where they use research to rationalize a plan of action can be a great first step toward professional competence. This book is designed to help you through the process of writing such a proposal.

The Six Parts of the Project Proposal

Though formats differ from organization to organization, there are always six basic parts to any good project proposal:

- Patron (the person who will fund your proposal)

- Population (the people who will benefit from it)

- Problem (the need or opportunity that your proposal addresses)
- Paradigm (the research rationale for your plan)
- Plan (the way you will address the problem)
- Price (the budget to implement your plan)

A good project proposal will always help a specific *population* to address a *problem* by developing a *paradigm*-based *plan* of action that stays within the *price* that your *patron* is willing to pay. Though formats will differ from place to place, all strong proposals will have these six basic elements. The difficult part of making a strong proposal is having all of the parts fit together into a coherent whole.

The Six P Formula can be used to organize both your written product and your writing process. The ideal process will follow the Six P's in order, more or less, focusing first on identifying a population to assist, a problem to solve, and a patron who would be willing to fund; then developing a paradigm through research that will help you design a well-justified plan of action; and once you have your plan you can develop your budget.

Patron

Who will fund your project? This is the person to whom you will literally address your proposal, and therefore the person whose name will be on your cover letter or memo. He or she will be your chief audience or reader. This is the person you most need to persuade. The patron could be your boss, the people at headquarters, a government group, or any public or private foundation. Like the "patrons" of the arts who paid for public and private projects during the Renaissance, your patron is the person whose hand controls the purse. Recognize that the interests of your patron must ultimately influence the proposal. If you choose a source that is most compatible with your approach, you will have fewer problems justifying your plan to them.

Population

Who will benefit from your project? These are the people who will directly or indirectly benefit from your proposal. In a business setting, they may be your customers or people in your organization. In the case of a scientific project, the population should be thought of as both the people in your field who want certain questions answered and the people in the world who might benefit from your research. In any case, a persuasive project proposal will have a well-thought-out human dimension. After all, why should a proposal be funded if no one but you will benefit from it?

Sometimes students choose a project proposal (such as any reform at their college) where they themselves are part of the population to be served. If you do that, it is especially important for you to remain objective. No one wants to fund a self-serving project, and until you can imagine other beneficiaries for your work (the larger population of students to be served) you will not be able to make a persuasive case for funding.

Problem

What instigates your project? This could be a theoretical question (in the case of scientific research), or an opportunity not to be missed (in the case of an entrepreneurial endeavor), or a persistent issue that needs remedying (in the narrowest sense of problem). All good proposals begin with a problem. If you find yourself beginning with a plan of action, then you have really jumped to conclusions. Before anyone should consider acting, after all, they need to be convinced that the problem is objectively real and that it needs to be addressed. You should first try to quantify or define your problem so that your patron can understand its scale, scope, and significance. In the case of a theoretical question, you will need to show how this question arose in prior research. In a business proposal, you will typically need to quantify the problem so that your patron can weigh the costs and benefits of action and inaction. Why does the problem even need to be addressed? Ultimately, you need to provide evidence to answer that question.

Paradigm

Why is your plan of action the best one available for addressing the problem? To answer "why" you need a research-based rationale that answers these two questions: *How do you know* that your plan will solve the problem? And *why* try to solve the problem *this way* rather than any number of other ways? A good research-based rationale will show that you have a consensus within your field that justifies your approach. It might also show that the plan you want to implement has strong precedents to suggest that it will succeed. This is basically what we mean by a "paradigm."

The way we use the term "paradigm" today has been greatly influenced by the work of Thomas Kuhn. In Kuhn's view, experiments form the basis of scientific knowledge by being what he called "exemplars," or models of how problems can be solved. The larger theory that explains why the models work, which he called the "disciplinary matrix," often comes much later. But the part and the whole are mutually dependent. When there is a consensus within a field of endeavor that this model and this matrix agree with each other, you have a paradigm. In terms of writing a persuasive project proposal, a paradigm can be either a model of success (or benchmark) that you think should be imitated or it can be a theoretical framework for understanding why your plan should succeed. The ideal paradigm will feature both an exemplar and a disciplinary matrix: it will have both a model of success and a theory of why that model succeeds. It will have both the part and the whole.

Paradigms describe the rhetorical and conceptual spaces that practitioners of any discipline generally follow. In the sciences and some areas of social research, paradigms are so commonly shared that a shorthand has developed for describing them, so that many paradigms can be summed up in a phrase that names a theory within a specific discipline: "integrated pest management" in agriculture, or "experiential learning" in education, or "ecological risk assessment" in environmental planning, or "the broken window theory" in sociology or law enforcement. These terms grew out of exemplary practices that became common knowledge within a field of endeavor.

If you want to develop a paradigm for your project, you might ask these questions: How have other people solved this problem or addressed this question in the past? What models of successful practice are available to give me ideas and help justify a plan or experiment? What theories or ideas might help me to develop a logical approach to this problem or to develop experimental procedures? How might language from my discipline help describe and understand the problem?

Once you have a paradigm, you will be able to construct your plan based on research. Without a paradigm you will be inventing your plan out of whole cloth with nothing but your own ethos to justify it, and that is not likely to take you very far with your patron.

Plan

Your plan might be a construction project, a training or education program, an experiment to test a hypothesis, a study to determine what course of action is best, or some other specific initiative. Since a good plan will have to grow organically out of the people, problem, and paradigm, it is generally not the first thing you will work on for the project. It has to be responsive to your research findings.

How you present your plan will depend upon your project, but you should strive to be as explicit as possible about all that will be involved. If you can find a way to visually organize this part of your proposal, it will help your reader to understand it better. If the project will take place in a series of steps, you might be able to set up a calendar showing the sequence of events. If the plan requires construction, you will probably want to draw a diagram of the thing you are going to build. But a good plan needs to look back at its problem and paradigm: it should detail the specific ways you are going to address the problem and suggest how it follows logically from your models and research.

Price

Once you have your plan in place, you will need to calculate a budget. Often, your budget is restricted before you begin your project, and you should recognize the ways that price can have a strong influence over choices you make in dealing with the other five P's. If you are making a case for overall long-term savings from your project, you may want to include those in your calculations. If the materials for your project can be broken down and detailed, then do so. Find out the price of the materials you need, either by contacting suppliers or looking up prices online. Talk to people who have done this sort of work before if you can. Use your judgment if you are not certain of costs, but try to be as realistic as possible.

Other Considerations

The Six P's are not an exhaustive list, but they should handle the critical issues you need to cover in any good proposal. We could add some other P's here, and I would like to mention two more, since they often come up: Partners and Politics.

By partners I mean the people who will help you achieve your goals yet who will not necessarily be benefiting from the project or providing funding for it. They might be other organizations or other people in your company. They can sometimes be very important to discuss in your proposal, since mentioning their support will show that you already have convinced other people that you have a good project idea.

By politics I mean the larger cultural, economic, legal, or political situation that may impact your proposal. As we know, projects that might gain support at one time or in one place will not gain support in another time or place. If your project runs counter to prevailing ideology, you may have a problem on your hands. For example, it would be politically difficult to get backing to promote the medical use of illegal drugs in a state with tough anti-drug laws; it would be difficult to organize a deer hunt to address a deer overpopulation issue in a community that is anti-hunting; it might be foolhardy to propose new accounting tricks in the wake of accounting scandals like Enron's; and costly projects will not be well received during times of fiscal difficulty. At the very least, you may need to give special attention to your rhetorical frame (that is, how you argue for your project) or you may need to adjust some of your assumptions to make your proposal more feasible given current realities (or "politics").

The Interdependence of the Six P's

You should imagine the Six P's spatially, as the parts of a coherent project that might come together in any temporal order. Making the Six P's fit together can sometimes feel like building a structure with six interlocking parts. The Six P's are completely interdependent entities, so choices in one area impact choices in other areas. You need to be open to revising the different parts of your project as it develops. How will decisions about the funding source (the addressee for your proposal) affect the way you approach the problem? How will the population to be served by the project impact the approach you might take?

For example, suppose your lab has expanded beyond its present capacity to give experimental space to all who need it. The people in charge of finding a solution to this problem will begin by asking themselves a number of questions:

- Patron: Where might we get money to solve this problem?
- Population: Who is most affected by the problem?
- Problem: What are some of the causes of the problem? What is its scope? What nonsubjective evidence do we have that there really is a problem?
- Paradigm: How have other labs succeeded in solving this problem? What innovative approaches (such as time sharing) have they used? What areas of knowledge can be brought to bear on the problem?

- Plan: What plans are feasible given current fiscal and political realities?
- Price: How much do you think you might be able to raise to fund your project? How much have similar projects cost?

If you find a patron willing to give you whatever money you need, then that will make it possible to build additional space. But if your funds are limited and you need to make do with the space available, then that will clearly change your approach. In the case of limited funds, you may need to make decisions about which researchers should have priority over others, and that will create a narrower population that needs extra assistance. We could go on and on, but clearly at each step of your project you should recognize how your choices can have cascading effects down the line.

The Six P's in Action

Because you are likely to make changes in your proposal at each step of its development, you should be prepared to revise your project as you go in order to make it more coherent. While not every project develops in a coherent step-by-step process following the order of the Six P's, they all need to put the Six P's together in a way that meshes. Some projects might actually begin with a plan ("I want to build a playground") and then work backwards in order to supply the paradigm, problem, and people that will create a unified project ("Who else needs a playground? Who else might benefit from the project and want to get involved? Who might fund it? How do we justify the project to the funding source?"). Some projects begin logically but then require extensive revision to resolve conflicts between various areas of the Six P's (as when the patron doesn't like the price).

A Rutgers computer science student taking one of our professional writing courses—let's call her Sandy—wanted to build a web page for a restaurant where she worked, but she didn't see how that web page could be used to improve business. She had begun with the plan ("I want to build a website"), and her dilemma was that she didn't see the problem to be solved or the paradigm to solve it. But that didn't mean she could not succeed in creating a fundable project proposal. She just had to go back to the beginning and ask some of the questions that had been skipped over in leaping to the plan of action.

Sandy already knew the funding source: the restaurant owners would pay to develop a good website. But she had not yet thought about the people to be served (Who are our customers? Do they even have access to the web?), or the problem (How could a website improve business? What opportunities are we missing out on by not having one?), or the paradigm to guide her (What principles or models of success might give us ideas?). Without answers to these questions, the website could very well become a waste of resources. She had to work on the Six P's.

A good project always depends on good research. And what Sandy most needed was a paradigm to guide her research.

In her initial writings about the project, Sandy had made the textbook distinction between "target marketing," which seeks to attract new consumers from a specific group, and "relationship marketing," which involves improving loyalty among the base of consumers who already use your product or service. She suggested that the Internet was probably more useful for relationship marketing than for target marketing because of how expensive and difficult it is to reach consumers who haven't already heard of your business. In fact, she recognized it might be easier to attract people to the website by using the restaurant than to attract people to the restaurant by using the website.

What Sandy did not recognize right away was that the term "relationship marketing" describes a researchable concept, literally a marketing paradigm. Sandy had stumbled upon the term in her initial research, but because she was not a marketing major (she was a computer science major, after all) it had not occurred to her that she could explore that concept further through more focused research and reading. To do that, she needed to look at resources in the marketing field and examples of relationship marketing in action. A brief stop at the library index ABI Inform (which indexes business sources and even offers full-text versions online) showed her that there was a wealth of source material within easy reach. A single search turned up almost 500 potential sources on "relationship marketing" alone. Though she found no examples of restaurants using the concept, she did discover quite a few service-sector models for using a website to build relationships with loyal customers. One of the best examples she found, described at length in one article, was of a dry cleaner that used a sophisticated website not only to communicate with customers but also to offer other services that helped to develop a sense of community around the establishment. The site even had a singles meeting page that allowed people going to this dry cleaner to connect with local singles, many of whom would post their pictures both online and in the lobby of the establishment.

You might say that Sandy's paradigm was supported by both the exemplar (or example) of the dry cleaner and the disciplinary matrix (or theory) of "relationship marketing," two things she knew nothing about when she began her research. There were other approaches she could have explored (for example, there is a large body of research on building a "virtual community," a term coined by Howard Rheingold). But the approach she found gave her what she needed. The idea Sandy ended up developing was quite creative and went beyond the things she had learned as a computer science major. In many ways, the project helped her to understand the human dimensions of her field.

After looking at how a number of other companies used their websites to develop relationships with customers, Sandy was able to synthesize an original yet proven plan for her workplace. She decided to work on developing a sense of community around the restaurant, so that even when customers were not there they could participate in the social life of the institution, developing a relationship with it like the patrons of the television bar "Cheers." To entice current customers of the restaurant to visit the website, she would offer them online coupons, on the model of a number of other establishments. Customers visiting the site could find out more about the staff, e-mail suggestions directly to the chef, check out the calendar of upcoming events, join the restaurant "listserv" to receive announcements and advertisements, or check out what was going on at one of the live chat rooms. In a business built on loyal customers, a website that helped build loyalty was a concept worth implementing.

It took research to lead the way.

Works Cited

Drucker, Peter F. "The Age of Social Transformation." *The Atlantic Monthly* 274.5 (November 1994): 53–80.

Kuhn, Thomas. *The Structure of Scientific Revolutions*. Chicago: University of Chicago Press, 1970.

Newspaper Exercise

Preparation

To participate in this exercise, you must read several newspapers and choose **one** article that could be the basis for a project proposal. The best articles will suggest a problem that you can imagine trying to address. The purpose of this exercise is *not* to get you to choose the topic that you will actually work on for this class (though it is possible that some of you might stumble upon your topic this way). Rather, this exercise is intended to get you to practice the process of project development. After reading Chapter One, take notes on the Six P's that you can imagine developing from the article you have chosen. This will be collected and used in class discussions.

In Class

Get into groups of four or five, and do the following:

1. Each of you, in turn, should present your article to the other group members. Describe the article and explain how this might make a good project idea. (About 10 minutes)

2. After the individual presentations, decide, as a group, which of them would make the most interesting basis for a project. (About 5 minutes)

3. Elect a group leader. This person does not have to be the one whose article was chosen, but he or she will present your ideas to the class. (About 5 minutes)

4. Once you have elected a group leader, begin developing a project idea based upon the article and on your own general knowledge. Obviously, to develop a strong project you would have to do research, but do the best you can. Use the following questions as a guide to discussion: (About 20 minutes)

    *   Patron. Who might fund your idea? Why would they fund it?

    *   Population. Who is affected by this problem? What specific population will your project serve?

    *   Problem. What is the basic problem or need your project will address? Why is it a problem? How could you illustrate the extent of this problem? What are the tangible effects of this issue upon the population?

    *   Paradigm. What disciplines (e.g., marketing, education, nutrition, medicine, etc.) might be useful in addressing this problem? What research would help?

    *   Plan. How might you address the problem you have identified? What is your plan of action? Who will carry out this plan?

    *   Price. What resources or assistance will you need? How much do you think your project might cost?

5. Present your ideas to the class and answer any questions. (5 minutes for each group)

Chapter 1 ■ The Project Proposal from the Ground Up

Six P's Exercise

Use the following form in a class exercise as directed by your instructor to analyze your project idea or the idea of someone else in your class.

Patron
Who would be willing to fund this project? Why would they want to fund it?

Population
Who does the problem affect? That is, who has a stake in seeing that there is a solution to the problem? Does your population have the same interests as the patron?

Problem
What are the main problems that need to be addressed? How could research shed light on these problems to emphasize their scale, scope, and significance? What sources of information about the problem would the patron find most persuasive?

Paradigm
What disciplines (e.g., computer science, marketing, education, psychology, etc.) might be useful in developing a disciplinary matrix for providing a rationale for action? Where might models of success be found to help shape the plan? What research would help?

Plan
What possible plans of action can you already imagine at this point? What plans are politically feasible? What would you need to know in order to develop a logical plan?

Price
How might your budget be limited? How much do you think the project might cost? How can that spending be justified?

Readings in Scientific and Technical Writing

Chapter 2

Each of the following readings was chosen to facilitate class discussion of the "paradigm" concept, which is central to understanding the social situation of research writing in scientific and technical fields. Without understanding the paradigm concept, you may have trouble conducting your research and successfully supporting your argument. You will also be missing a key concept of our times that has had a broad impact on thinking in many disciplines.

Without being familiar with the paradigm concept, you might not think of how important it is to review the tradition in your field to support your innovative work, as Thomas Kuhn discusses. Paradigms are central to the work that Kuhn calls "normal science," but we must remember that paradigms both shape and are shaped by cultural and technological forces, as Ian Parker's history of PowerPoint's hegemony suggests. Paradigms support innovative research, as the new "germ theory" described by Judith Hooper has done. But they can also lead to overly dogmatic applications, as she also suggests. An alternative path toward innovation that avoids such dogmatism is to adapt paradigms from other fields to your own. Malcolm Gladwell's essay on "designs for working" demonstrates that the paradigms that guide our plans can come from theoretical sources we might not expect to find useful. Who would have thought, after all, that the ideas of sociologist and urban planner Jane Jacobs would be so useful in designing office space? Paradigms, which are the product of consensus, are ultimately social products, and as such they are rarely without critics, because they can dramatically affect people's lives. Views on "hot" topics such as global warming, the subject of Daniel Sarewitz and Roger Pielke, Jr.'s article, are a case in point. Paradigms make winners and losers: those whose work is funded and those whose work is not, for example. Being able to situate your work within such traditions and use them to innovate is the key to being a strong research writer in scientific and technical fields.

References

Gladwell, M. (2000, December 11). Designs for working. *New Yorker*, 60–70.

Hooper, J. (1999, February). A new germ theory. *Atlantic Monthly*, 41–53.

Kuhn, T. (1959) The essential tension: tradition and innovation in scientific research. In C. W. Taylor (Ed.), *The University of Utah research conference on the identification of scientific talent.* Salt Lake City: University of Utah Press.

Parker, Ian. (2001, May 28) Absolute PowerPoint. *New Yorker*, 78+.

Sarewitz, D. & Pielke, R. Jr. (2000, July). Breaking the global warming gridlock. *Atlantic Monthly*, 55–64.

The Essential Tension

Tradition and Innovation in Scientific Research

Thomas Kuhn

■ ■ ■

Thomas Kuhn is very well known in the history of science, and his book *The Structure of Scientific Revolutions* remains one of the most cited academic works of the last century. Kuhn's use of the term "paradigm" had a lasting impact on the language, twisting a term that had previously meant "model" or "example" so that it signified a consensus within a field of research (among other things), which can sometimes very quickly follow innovations (what Kuhn termed "paradigm shifts"). The following selection from his work is a conference talk he presented at the University of Utah to academics interested in creativity and the development of scientific talent. It retains much of the flavor of a public presentation and, as such, makes many of Kuhn's concepts accessible to a general audience.

I am grateful for the invitation to participate in this important conference, and I interpret it as evidence that students of creativity themselves possess the sensitivity to divergent approaches that they seek to identify in others. But I am not altogether sanguine about the outcome of your experiment with me. As most of you already know, I am no psychologist, but rather an ex-physicist now working in the history of science. Probably my concern is no less with creativity than your own, but my goals, my techniques, and my sources of evidence are so very different from yours that I am far from sure how much we do, or even should, have to say to each other. These reservations imply no apology: rather they hint at my central thesis. In the sciences, as I shall suggest below, it is often better to do one's best with the tools at hand than to pause for contemplation of divergent approaches.

If a person of my background and interests has anything relevant to suggest to this conference, it will not be about your central concerns, the creative personality and its early identification. But implicit in the numerous working papers distributed to participants in this conference is an image of the scientific process and of the scientist; that image almost certainly conditions many of the experiments you try as well as the conclusions you draw; and about it the physicist-historian may well have something to say. I shall restrict my attention to one aspect of this image—an aspect epitomized as follows in one of the working papers: The basic scientist "must lack prejudice to a degree where he can look at the most 'self-evident' facts or concepts without necessarily accepting them, and, conversely, allow his imagination to play with the most unlikely possibilities" (Selye, 1959). In the more technical language supplied by other working papers (Getzels and Jackson), this aspect of the image recurs as an emphasis upon "divergent thinking, . . . the freedom to go off in different directions, . . . rejecting the old solutions and striking out in some new direction."

I do not at all doubt that this description of "divergent thinking" and the concomitant search for those able to do it are entirely proper. Some divergence characterizes all scientific work, and gigantic divergences lie at the core of the most significant episodes in scientific development. But both my own experiences in scientific research and my reading of the history of the sciences lead me to wonder whether flexibility and open-mindedness have not been too exclusively emphasized as the characteristics requisite for basic research. I shall therefore suggest below that something like "convergent thinking" is just as essential to scientific advances as is divergent. Since these two modes of thought are inevitably in conflict, it will follow that the ability to support a tension that can occasionally become almost unbearable is one of the prime requisites for the very best sort of scientific research.

I am elsewhere studying these points more historically, with emphasis on the importance to scientific development of "revolutions." These are episodes—exemplified in their most extreme and readily recognized form by the advent of Copernicanism, Darwinism, or Einsteinianism—in which a scientific community abandons one time-honored way of regarding the world and of pursuing science in favor of some other, usually incompatible, approach to its discipline. I have argued in the draft that the historian constantly encounters many far smaller but structurally similar revolutionary episodes and that they are central to scientific advance. Contrary to a prevalent impression, most new discoveries and theories in the sciences are not merely additions to the existing stockpile of scientific knowledge. To assimilate them the scientist must usually rearrange the intellectual and manipulative equipment he has previously relied upon, discarding some elements of his prior belief and practice while finding new significances in and new relationships between many others. Because the old must be revalued and reordered when assimilating the new, discovery and invention in the sciences are usually intrinsically revolutionary. Therefore, they do demand just that flexibility and open-mindedness that characterize, or indeed define, the divergent thinker. Let us henceforth take for granted the need for these characteristics. Unless many scientists possessed them to a marked degree, there would be no scientific revolutions and very little scientific advance.

Yet flexibility is not enough, and what remains is not obviously compatible with it. Drawing from various fragments of a project still in progress, I must now emphasize that revolutions are but one of two complementary aspects of scientific advance. Almost none of the research undertaken by even the greatest scientists is designed to be revolutionary, and very little of it has any such effect. On the contrary, normal research, even the best of it, is a highly convergent activity based firmly upon settled consensus acquired from scientific education and reinforced by subsequent life in the profession. Typically, to be sure, this convergent or consensus-bound research ultimately results in revolution. Then, traditional techniques and beliefs are abandoned and replaced by new ones. But revolutionary shifts of a scientific tradition are relatively rare, and extended periods of convergent research are the necessary preliminary to them. As I shall indicate below, only investigations break that tradition and give rise to a new one. That is why I speak of an "essential tension" implicit in scientific research. To do his job the scientist must undertake a complex set of intellectual and manipulative commitments. Yet his claim to fame, if he has the talent and good luck to gain one, may finally rest upon his ability to abandon this net of commitments in favor of another of his own invention. Very often the successful scientist must simultaneously display the characteristics of the traditionalist and of the iconoclast.

The multiple historical examples upon which any full documentation of these points must depend are prohibited by the time limitations of the conference. But another approach will introduce you to at least part of what I have in mind—an examination of the nature of education in the natural sciences. One of the working papers for this conference (Getzels and Jackson) quotes Guilford's very apt description of scientific education as follows: "[It] has emphasized abilities in the areas of convergent thinking and evaluation, often at the expense of development in the area of divergent thinking. We have attempted to teach students how to arrive at 'correct' answers that our civilization has taught us are correct. . . . Outside the arts [and I should include most of the social sciences] we have generally discouraged the development of divergent-thinking abilities, unintentionally." That characterization seems to me eminently just, but I wonder whether it is equally just to deplore the product that results. Without defending plain bad teaching, and granting that in this country the trend to convergent thinking in all education may have proceeded entirely too far, we may nevertheless recognize that a rigorous training in convergent thought has been intrinsic to the sciences almost from their origin. I suggest that they could not have achieved their present state or status without it.

Let me try briefly to epitomize the nature of education in the natural sciences, ignoring the many significant yet minor differences between the various sciences and between the approaches of different educational institutions. The single most striking feature of this education is that, to an extent totally unknown in other creative fields, it is conducted entirely through textbooks. Typically, undergraduate and graduate students of chemistry, physics, astronomy, geology, or biology acquire the substance of their fields from books written especially for students. Until they are ready, or even nearly ready, to commence work on their own dissertations, they are neither asked to attempt trial research projects nor exposed to the immediate products of research done by others, that is, to the professional communications that scientists write for each other. There are no collections of "readings" in the natural sciences. Nor are science students encouraged to read the historical classics of their fields—works in which they might discover other ways of regarding the problems discussed in their textbooks, but in which they would also meet problems, concepts, and standards of solution that their future professions have long since discarded and replaced.

In contrast, the various textbooks that the students does encounter display different subject matters, rather than, as in many of the social sciences, exemplifying different approaches to a single problem field. Even books that compete for adoption in a single course differ mainly in level and in pedagogic detail, not in substance or conceptual structure. Last, but most important of all, is the characteristic technique of textbook presentation. Except in their occasional introductions, science textbooks do not describe the sorts of problems that the professional may be asked to solve and the variety of techniques available for their solution. Rather, these books exhibit concrete problem solutions that the profession has come to accept as paradigms, and they then ask the student, either with a pencil and paper or in the laboratory, to solve for himself problems very closely related in both method and substance to those through which the textbook or the accompanying lecture has led him. Nothing could be better calculated to produce "mental sets" or *Einstelleungen*. Only in their most elementary courses do other academic fields offer as much as a partial parallel.

Even the most faintly liberal educational theory must view this pedagogic technique as anathema. Students, we would all agree, must begin by learning a good deal of what is already known, but we also insist that education give them vastly more. They must, we say, learn to recognize and evaluate problems to which no unequivocal solution has yet been given; they must be supplied with an arsenal of techniques for approaching these future problems; and they must learn to judge the relevance of these techniques and to evaluate the possibly partial solutions that they can provide. In many respects these attitudes toward education seem to me entirely right, and yet we must recognize two things about them. First, education in the natural sciences seems to have been totally unaffected by their existence. It remains a dogmatic initiation in a pre-established tradition that the student is not equipped to evaluate. Second, at least in the period when it was followed by a term in an apprenticeship relation, this technique of exclusive exposure to a rigid tradition has been immensely productive of the most consequential sorts of innovations.

I shall shortly inquire about the pattern of scientific practice that grows out of this educational initiation and will then attempt to say why that pattern proves quite so successful. But first, an historical excursion will reinforce what has just been said and prepare the way for what is to follow. I should like to suggest that the various fields of natural science have not always been characterized by rigid education in exclusive paradigms, but that each of them acquired something like that technique at precisely the point when the field began to make rapid and systematic progress. If one asks about the origin of our contemporary knowledge of chemical composition, of earthquakes, of biological reproduction, of motion through space, or of any other subject matter known to the natural sciences one immediately encounters a characteristic pattern that I shall here illustrate with a single example.

Today, physics textbooks tell us that light exhibits some properties of a wave and some of a particle: both textbook problems and research problems are designed accordingly. But both this view and these textbooks are products of an early twentieth-century revolution. (One characteristic of scientific revolutions is that they call for the rewriting of science textbooks.) For more than half a century before 1900, the books employed in scientific education had been equally unequivocal in stating that light was wave motion. Under those circumstances scientists worked on somewhat different problems and sometimes embraced rather different sorts of solutions to them. The nineteenth-century textbook tradition does not, however, mark the beginning of our subject matter. Throughout the eighteenth century and into the early nineteenth, Newton's *Opticks* and the other books from which men learned science taught almost all students that light was particles, and research guided by this tradition was again different from that which succeeded it. Ignoring a variety of subsidiary changes within these three successive traditions, we may therefore say that our views derive historically from Newton's views by way of two revolutions in optical thought, each of which replaced one tradition of convergent research with another. If we make appropriate allowances for changes in the locus and materials of scientific education, we may say that each of these three traditions was embodied in the sort of education by exposure to unequivocal paradigms that I briefly epitomized above. Since Newton, education and research in physical optics have normally been highly convergent.

The history of theories of light does not, however, begin with Newton. If we ask about knowledge in the field before his time, we encounter a significantly different pattern—a pattern still familiar in the arts and in some social sciences, but one that has largely disappeared in the natural sciences. From remote antiquity until the end of the seventeenth century there was no single set of paradigms for the study of physical optics. Instead, many men advanced a large number of different views about the nature of light. Some of these views found few adherents, but a number of them gave rise to continuing schools of optical thought. Although the historian can note the emergence of new points of view as well as changes in the relative popularity of older ones, there was never anything resembling consensus. As a result, a new man entering the field was inevitably exposed to a variety of conflicting viewpoints; he was forced to examine the evidence for each, and there always was good evidence. The fact that he made a choice and conducted himself accordingly could not entirely prevent his awareness of other possibilities. This earlier mode of education was obviously more suited to produce a scientist without prejudice, alert to novel phenomena, and flexible in his approach to his field. On the other hand, one can scarcely escape the impression that, during the period characterized by this more liberal educational practice, physical optics made very little progress.

The pre-consensus (we might here call it the divergent) phase in the development of physical optics is, I believe, duplicated in the history of all other scientific specialties, excepting only those that were born by the subdivision and recombination of pre-existing disciplines. In some fields, like mathematics and astronomy, the first firm consensus is prehistoric. In others, like dynamics, geometric optics, and parts of physiology, the paradigms that produced a first consensus date from classical antiquity. Most other natural sciences, though their problems were often discussed in antiquity, did not achieve a first consensus until after the Renaissance. In physical optics, as we have seen, the first firm consensus dates only from the end of the seventeenth century; in electricity, chemistry, and the study of heat, it dates from the eighteenth; while in geology and the non-taxonomic parts of biology no very real consensus developed until after the first third of the nineteenth century. This century appears to be characterized by the emergence of a first consensus in parts of a few of the social sciences.

In all the fields named above, important work was done before the achievement of the maturity produced by consensus. Neither the nature nor the timing of the first consensus in these fields can be understood without a careful examination of both the intellectual and the manipulative techniques developed before the existence of unique paradigms. But the transition to maturity is not less significant because individuals practiced science before it occurred. On the contrary, history strongly suggests that, though one can practice science—as one does philosophy or art or political science—without a firm consensus, this more flexible practice will not produce the pattern of rapid consequential scientific advance to which recent centuries have accustomed us. In that pattern, development occurs from one consensus to another, and alternate approaches are not ordinarily in competition. Except under quite special conditions, the practitioner of a mature science does not pause to examine divergent modes of explanation or experimentation.

I shall shortly ask how this can be so—how a firm orientation toward an apparently unique tradition can be compatible with the practice of the disciplines most noted for the persistent production of novel ideas and techniques. But it will help first to ask what the education that so successfully transmits such a tradition leaves to be done. What can a scientist working within a deeply rooted tradition and little trained in the perception of significant alternatives hope to do in his professional career? Once again limits of time force me to drastic simplification, but the following remarks will at least suggest a position that I am sure can be documented in detail.

In pure or basic science—that somewhat ephemeral category of research undertaken by men whose most immediate goal is to increase understanding rather than control of nature—the characteristic problems are almost always repetitions, with minor modifications, of problems that have been undertaken and partially resolved before. For example, much of the research undertaken within a scientific tradition is an attempt to adjust existing theory or existing observation in order to bring the two into closer and closer agreement. The constant examination of atomic and molecular spectra during the years since the birth of wave mechanics, together with the design of theoretical approximations for the prediction of complex spectra, provides one important instance of this typical sort of work. Another was provided by the remarks about the eighteenth-century development of Newtonian dynamics in [his paper on measurement.] The attempt to make existing theory and observation conform more closely is not, of course, the only standard sort of research problem in the basic sciences. The development of chemical thermodynamics or the continuing attempts to unravel organic structure illustrate another type—the extension of existing theory to areas that it is expected to cover but in which it has never before been tried. In addition, to mention a third common sort of research problem, many scientists constantly collect the concrete data (e.g.: atomic wrights, nuclear moments) required for the application and extension of existing theory.

These are normal research projects in the basic sciences, and they illustrate the sorts of work on which all scientists, even the greatest, spend most of their professional lives and on which many spend all. Clearly their pursuit is neither intended nor likely to produce fundamental discoveries or revolutionary changes in scientific theory. Only if the validity of the contemporary scientific tradition is assumed do these problems make much theoretical or any practical sense. The man who suspected the existence of a totally new type of phenomenon or who had basic doubts about the validity of existing theory would not think problems so closely modeled on textbook paradigms worth undertaking. It follows that the man who does undertake a problems of this sort—and that means all scientists at most times—aims to elucidate the scientific tradition in which he was raised rather than to change it. Furthermore, the fascination of his work lies in the difficulties of elucidation rather than in any surprises that the work is likely to produce. Under normal conditions the research scientist is not an innovator. Under normal conditions the research scientist is not an innovator but a solver of puzzles, and the puzzles upon which he concentrates are just those that he believes can be both stated and solved within the existing scientific tradition.

Yet—and this is the point—the ultimate effect of this tradition-bound work has invariably been to change the tradition. Again and again the continuing attempt to elucidate a currently received tradition has at least produced one of those shifts in fundamental theory, in problem field, and in scientific standards to which I previously referred as scientific revolutions. At least for the scientific community as a whole, work within a well-defined and deeply ingrained tradition seems more productive of tradition-shattering novelties than work in which no similarly convergent standards are involved.

How can this be so? I think it is because no other sort of work is nearly so well suited to isolate for continuing and concentrated attention those loci of trouble or causes of crisis upon whose recognition the most fundamental advances in basic science depend.

As I have indicated in the first of my working papers, new theories and, to an increasing extent, novel discoveries in the mature sciences are not born de novo. On the contrary, they emerge from old theories and within a matrix of old beliefs about the phenomena that the world does and does not contain. Ordinarily such novelties are far too esoteric and recondite to be noted by the man without a great deal of scientific training. And even the man with considerable training can seldom afford simply to go out and look for them, let us say by exploring those areas in which existing data and theory have failed to produce understanding. Even in a mature science there are always far too many such areas, areas in which no existing paradigms seem obviously to apply and for whose exploration few tools and standards are available. More likely than not the scientist who ventured into them, relying merely upon his receptivity to new phenomena and his flexibility to new patterns of organization, would get nowhere at all. He would rather return his science to its pre-consensus or natural history phase.

Instead, the practitioner of a mature science, from the beginning of his doctoral research, continues to work in the regions for which the paradigms derived from his education and from the research of his contemporaries seem adequate. He tries, that is, to elucidate topographical detail on a map whose main outlines are available in advance, and he hopes—if he is wise enough to recognizes the nature of his field—that he will some day undertake a problem in which the anticipated does not occur, a problem that goes wrong in ways suggestive of a fundamental weakness in the paradigm itself. In the mature sciences the prelude to much discovery and to all novel theory is not ignorance, but the recognition that something has gone wrong with existing knowledge and beliefs.

What I have said so far may indicate that it is sufficient for the productive scientist to adopt existing theory as a lightly held tentative hypothesis, employ it *faut de mieux* in order to get a start in his research, and then abandon it as soon as it leads him to a trouble spot, a point at which something has gone wrong. But though the ability to recognize trouble when confronted by it is surely a requisite for scientific advance, trouble must not be too easily recognized. The scientist requires a thoroughgoing commitment to the tradition with which, if he is fully successful, he will break. In part this commitment is demanded by the nature of the problems the scientist normally undertakes. These, as we have seen, are usually esoteric puzzles whose challenge lies less in the information disclosed by their solutions (all but its details are often known in advance) than in the difficulties of technique to be surmounted in providing any solution at all. Problems of this sort are undertaken only by men assured that there is a solution that ingenuity can disclose, and only current theory could possibly provide assurance of that sort. That theory alone gives meaning to most of the problems of normal research. To doubt it is often to doubt that the complex technical puzzles which constitute normal research have any solutions at all. Who, for example, would have developed the elaborate mathematical techniques required for the study of the effects of interplanetary attractions upon basic Keplerian orbits if he had not assumed that Newtonian dynamics, applied to the planets then known, would explain the last details of astronomical observation? But without that assurance, how would Neptune have been discovered and the list of planets changed?

In addition, there are pressing practical reasons for commitment. Every research problem confronts the scientist with anomalies whose sources he cannot quite identify. His theories and observations never quite agree; successive observations never yield quite the same results; his experiments have both theoretical and phenomenological by-products that it would take another research project to unravel. Each of these anomalies or incompletely understood phenomena could conceivably be the clue to a fundamental innovation in scientific theory or technique, but the man who pauses to examine them one by one never completes his first project. Reports of effective research repeatedly imply that all but the most striking and central discrepancies could be taken care of by current theory if only there were time to take them on. The men who make these reports find most discrepancies trivial or uninteresting, an evaluation that they can ordinarily base only upon their faith in current theory. Without that faith their work would be wasteful of time and talent.

Besides, lack of commitment too often results in the scientist's undertaking problems that he has little chance of solving. Pursuit of an anomaly is fruitful only if the anomaly is more than non-trivial. Having discovered it, the scientist's first efforts and those of his profession are to do what nuclear physicists are now doing. They strive to generalize the anomaly, to discover other and more revealing manifestations of the same effect, to give it structure by examining its complex interrelationships with phenomena they still feel they understand. Very few anomalies are susceptible to this sort of treatment. To be so they must be in explicit and unequivocal conflict with some structurally central tenet of current scientific belief. Therefore, their recognition and evaluation once again depend upon a firm commitment to the contemporary scientific tradition.

This central role of an elaborate and often esoteric tradition is what I have principally had in mind when speaking of the essential tension in scientific research. I do not doubt that the scientist must be, at least potentially, an innovator, that he must possess mental flexibility, and that he must be prepared to recognize troubles where they exist. That much of the popular stereotype is surely correct, and it is important accordingly to search for indices of the corresponding personality characteristics. But what is no part of our stereotype and what appears to need careful integration with it is the other face of this same coin. We are, I think, more likely fully to exploit our potential scientific talent if we recognize the extent to which the basic scientist must also be a firm traditionalist, or, if I am using your vocabulary at all correctly, a convergent thinker. Most important of all, we must seek to understand how these two superficially discordant modes of problem solving can be reconciled both within the individual and within the group.

Everything said above needs both elaboration and documentation. Very likely some of it will change in the process. This paper is a report on work in progress. But, though I insist that much of it is tentative and all of it incomplete, I still hope that the paper has indicated why an educational system best described as an initiation into an unequivocal tradition should be thoroughly compatible with successful scientific work. And I hope, in addition, to have made plausible the historical thesis that no part of science has progressed very far or very rapidly before this convergent education and correspondingly convergent normal practice became possible. Finally, though it is beyond my competence to derive personality correlates from this view of scientific development, I hope to have made meaningful the view that the productive scientist must be a traditionalist who enjoys playing intricate games by pre-established rules in order to be a successful innovator who discovers new rules and new pieces with which to play them.

As first planned, my paper was to have ended at this point. But work on it, against the background supplied by the working papers distributed to conference participants, has suggested the need for a postscript. Let me therefore briefly try to eliminate a likely ground of misunderstanding and simultaneously suggest a problem that urgently needs a great deal of investigation.

Everything said above was intended to apply strictly only to basic science, an enterprise whose practitioners have ordinarily been relatively free to choose their own problems. Characteristically, as I have indicated, these problems have been selected in areas where paradigms were clearly applicable but where exciting puzzles remained about how to apply them and how to make nature conform to the results of the application. Clearly the inventor and applied scientist are not generally free to choose puzzles of this sort. The problems among which they may choose are likely to be largely determined by social, economic, or military circumstances external to the sciences. Often the decision to seek a cure for a virulent disease, a new source of household illumination, or an alloy able to withstand the intense heat of rocket engines must be made with little reference to the state of the relevant science. It is, I think, by no means clear that the personality characteristics requisite for pre-eminence in this more immediately practical sort of work are altogether the same as those required for a great achievement in basic science. History indicates that only a few individuals, most of whom worked in readily demarcated areas, have achieved eminence in both.

I am by no means clear where this suggestion leads us. The troublesome distinctions between basic research, applied research, and invention need far more investigation. Nevertheless, it seems likely, for example, that the applied scientist, to whose problems no scientific paradigm need be fully relevant, may profit by a far broader and less rigid education than that to which the pure scientist has characteristically been exposed. Certainly there are many episodes in the history of technology in which lack of more than the most rudimentary scientific education has proved to be an immense help. This group scarcely needs to be reminded that Edison's electric light was produced in the face of unanimous scientific opinion that the arc light could not be "subdivided," and there are many other episodes of this sort.

This must not suggest, however, that mere differences in education will transform the applied scientist into a basic scientist or vice versa. One could at least argue that Edison's personality, ideal for the inventor and perhaps also for the "oddball" in applied science, barred him from fundamental achievements in the basic sciences. He himself expressed great scorn for scientists and thought of them as wooly-headed people to be hired when needed. But this did not prevent his occasionally arriving at the most sweeping and irresponsible scientific theories of his own. (The pattern recurs in the early history of electrical technology: both Tesla and Gramme advanced absurd cosmic schemes that they thought deserved to replace the current scientific knowledge of their day.) Episodes like this reinforce an impression that the personality requisites of the pure scientist and of the inventor may be quite different, perhaps with those of the applied scientist lying somewhere between.

Is there a further conclusion to be drawn from all this? One speculative thought forces itself upon me. If I read the working papers correctly, they suggest that most of you are really in search of the *inventive* personality, a sort of person who does emphasize divergent thinking but whom the United States has already produced in abundance. In the process you may be ignoring certain of the essential requisites of the basic scientist, a rather different sort of person, to whose ranks America's contributions have as yet been notoriously sparse. Since most of you are, in fact, Americans, this correlation may not be entirely coincidental.

Chapter 2 ■ The Essential Tension

Questions for Discussion

1. Kuhn uses the term "paradigm" many times in his talk without pausing to define it. Based on your reading of how he uses the term, how would you define a "paradigm"?

   As part of the process of developing a definition, get into groups of three and try, as a group, to draw a picture of a paradigm. Of course, the word describes an abstract concept and so there can be no actual picture of a paradigm. But in drawing a picture you will be able to objectify it and thus make it easier to describe in your own words. Give it a try. What does a paradigm look like? How might you describe it metaphorically?

2. Students often say that they are prevented from being original by the requirement that they use research to provide a rationale for their plan. After all, they ask, where is the room for pure invention if everything needs to be supported by research into what other people have done? How can new ideas come out of old ones?

How might Kuhn respond to these questions? According to Kuhn, how does tradition actually *support* original research? What evidence does Kuhn use to support this claim? By what logic does consensus support rapid progress within a research field?

3. Kuhn was speaking at a conference whose main objective was to help foster scientific talent and creativity. How might colleges do more to encourage their students to pursue original ideas? What are some of the problems that Kuhn identifies in our typical science education system, and how might they be remedied?

4. Kuhn several times points out that the arts and social sciences have multiple paradigms while the sciences tend toward a single unifying paradigm in each field. What reasons does he suggest for this disparity? What does this suggest about the potential difficulties of people working in the arts and social sciences? Should scientists necessarily adopt more of a liberal arts model? What would be gained and lost by doing so?

5. Within what specific field of study is your own project situated? What paradigm governs that field? How does that paradigm inform your own work? Would you say that there is a "mini-paradigm" within your specific sub-field of research? How would you describe that paradigm? Does it derive more from a theory or from a specific experiment or model?

Designs for Working

Why your bosses want to turn your new office into Greenwich Village.

Malcolm Gladwell

— ▪ — ▪ — ▪ —

In the early nineteen-sixties, Jane Jacobs lived on Hudson Street, in Greenwich Village, near the intersection of Eighth Avenue and Bleecker Street. It was then, as now, a charming district of nineteenth-century tenements and town houses, bars and shops, laid out over an irregular grid, and Jacobs loved the neighborhood. In her 1961 masterpiece, "The Death and Life of Great American Cities," she rhapsodized about the White Horse Tavern down the block, home to Irish longshoremen and writers and intellectuals—a place where, on a winter's night, as "the doors open, a solid wave of conversation and animation surges out and hits you." Her Hudson Street had Mr. Slube, at the cigar store, and Mr. Lacey, the locksmith, and Bernie, the candy-store owner, who, in the course of a typical day, supervised the children crossing the street, lent an umbrella or a dollar to a customer, held on to some keys or packages for people in the neighborhood, and "lectured two youngsters who asked for cigarettes." The street had "bundles and packages, zigzagging from the drug store to the fruit stand and back over to the butcher's," and "teenagers, all dressed up, are pausing to ask if their slips show or their collars look right." It was, she said, an urban ballet. The miracle of Hudson Street, according to Jacobs, was created by the particular configuration of the streets and buildings of the neighborhood. Jacobs argued that when a neighborhood is oriented toward the street, when sidewalks are used for socializing and play and commerce, the users of that street are transformed by the resulting stimulation: they form relationships and casual contacts they would never have otherwise. The West Village, she pointed out, was blessed with a mixture of houses and apartments and shops and offices and industry, which meant that there were always people "outdoors on different schedules and... in the place for different purposes." It had short blocks, and short blocks create the greatest variety in foot traffic. It had lots of old buildings, and old buildings have the low rents that permit individualized and creative uses. And, most of all, it had people, cheek by jowl, from every conceivable walk of life. Sparely populated suburbs may look appealing, she said, but without an active sidewalk life, without the frequent, serendipitous interactions of many different people, "there is no public acquaintanceship, no foundation of public trust, no cross-connections with the necessary people—and no practice or ease in applying the most ordinary techniques of city public life at lowly levels."

Jane Jacobs did not win the battle she set out to fight. The West Village remains an anomaly. Most developers did not want to build the kind of community Jacobs talked about, and most Americans didn't want to live in one. To reread "Death and Life" today, however, is to be struck by how the intervening years have given her arguments a new and unexpected relevance. Who, after all, has a direct interest in creating diverse, vital spaces that foster creativity and serendipity? Employers do. On the fortieth anniversary of its publication, "Death and Life" has been reborn as a primer on workplace design.

The parallels between neighborhoods and offices are striking. There was a time, for instance, when companies put their most valued employees in palatial offices, with potted plants in the corner, and secretaries out front, guarding access. Those offices were suburbs—gated communities, in fact—and many companies came to realize that if their best employees were isolated in suburbs they would be deprived of public acquaintanceship, the foundations of public trust, and cross-connections with the necessary people. In the eighties and early nineties, the fashion in corporate America was to follow what designers called "universal planning"—rows of identical cubicles, which resembled nothing so much as a Levittown. Today, universal planning has fallen out of favor, for the same reason that the postwar suburbs like Levittown did: to thrive, an office space must have a diversity of uses—it must have the workplace equivalent of houses and apartments and shops and industry.

If you visit the technology companies of Silicon Valley, or the media companies of Manhattan, or any of the firms that self-consciously identify themselves with the New Economy, you'll find that secluded private offices have been replaced by busy public spaces, open-plan areas without walls, executives next to the newest hires. The hush of the traditional office has been supplanted by something much closer to the noisy, bustling ballet of Hudson Street. Forty years ago, people lived in neighborhoods like the West Village and went to work in the equivalent of suburbs. Now, in one of the odd reversals that mark the current economy, they live in suburbs and, increasingly, go to work in the equivalent of the West Village.

The office used to be imagined as a place where employees punch clocks and bosses roam the halls like high-school principals, looking for miscreants. But when employees sit chained to their desks, quietly and industriously going about their business, an office is not functioning as it should. That's because innovation—the heart of the knowledge economy—is fundamentally social. Ideas arise as much out of casual conversations as they do out of formal meetings. More precisely, as one study after another has demonstrated, the best ideas in any workplace arise out of casual contacts among different groups within the same company. If you are designing widgets for Acme.com, for instance, it is unlikely that a breakthrough idea is going to come from someone else on the widget team: after all, the other team members are as blinkered by the day-to-day demands of dealing with the existing product as you are. Someone from outside Acme.com—your old engineering professor, or a guy you used to work with at Apex.com—isn't going to be that helpful, either. A person like that doesn't know enough about Acme's widgets to have a truly useful idea. The most useful insights are likely to come from someone in customer service, who hears firsthand what widget customers have to say, or from someone in marketing, who has wrestled with the problem of how to explain widgets to new users, or from someone who used to work on widgets a few years back and whose work on another Acme product has given him a fresh perspective. Innovation comes from the interactions of people at a comfortable distance from one another, neither too close nor too far. This is why—quite apart from the matter of logistics and efficiency—companies have offices to begin with. They go to the trouble of gathering their employees under one roof because they want the widget designers to bump into the people in marketing and the people in customer service and the guy who moved to another department a few years back.

The catch is that getting people in an office to bump into people from another department is not so easy as it looks. In the sixties and seventies, a researcher at M.I.T. named Thomas Allen conducted a decade-long study of the way in which engineers communicated in research-and-development laboratories. Allen found that the likelihood that any two people will communicate drops off dramatically as the distance between their desks increases: we are four times as likely to communicate with someone who sits six feet away from us as we are with someone who sits sixty feet away. And people seated more than seventy-five feet apart hardly talk at all.

Allen's second finding was even more disturbing. When the engineers weren't talking to those in their immediate vicinity, many of them spent their time talking to people *outside* their company—to their old computer-science professor or the guy they used to work with at Apple. He concluded that it was actually easier to make the outside call than to walk across the room. If you constantly ask for advice or guidance from people inside your organization, after all, you risk losing prestige. Your colleagues might think you are incompetent. The people you keep asking for advice might get annoyed at you. Calling an outsider avoids these problems. "The engineer can easily excuse his lack of knowledge by pretending to be an `expert in something else' who needs some help in `broadening into this new area,'" Allen wrote. He did his study in the days before E-mail and the Internet, but the advent of digital communication has made these problems worse. Allen's engineers were far too willing to go outside the company for advice and new ideas. E-mail makes it even easier to talk to people outside the company.

The task of the office, then, is to invite a particular kind of social interaction—the casual, nonthreatening encounter that makes it easy for relative strangers to talk to each other. Offices need the sort of social milieu that Jane Jacobs found on the sidewalks of the West Village. "It is possible in a city street neighborhood to know all kinds of people without unwelcome entanglements, without boredom, necessity for excuses, explanations, fears of giving offense, embarrassments respecting impositions or commitments, and all such paraphernalia of obligations which can accompany less limited relationships," Jacobs wrote. If you substitute" office" for "city street neighborhood," that sentence becomes the perfect statement of what the modern employer wants from the workplace.

 Imagine a classic big-city office tower, with a floor plate of a hundred and eighty feet by a hundred and eighty feet. The center part of every floor is given over to the guts of the building: elevators, bathrooms, electrical, and plumbing systems. Around the core are cubicles and interior offices, for support staff and lower management. And around the edges of the floor, against the windows, are rows of offices for senior staff, each room perhaps two hundred or two hundred and fifty square feet. The best research about office communication tells us that there is almost no worse way to lay out an office. The executive in one corner office will seldom bump into any other executive in a corner office. Indeed, stringing the exterior offices out along the windows guarantees that there will be very few people within the critical sixty-foot radius of those offices. To maximize the amount of contact among employees, you really ought to put the most valuable staff members in the center of the room, where the highest number of people can be within their orbit. Or, even better, put all places where people tend to congregate—the public areas—in the center, so they can draw from as many disparate parts of the company as possible. Is it any wonder that creative firms often prefer loft-style buildings, which have usable centers?

Another way to increase communication is to have as few private offices as possible. The idea is to exchange private space for public space, just as in the West Village, where residents agree to live in tiny apartments in exchange for a wealth of nearby cafés and stores and bars and parks. The West Village forces its residents outdoors. Few people, for example, have a washer and dryer in their apartment, and so even laundry is necessarily a social event: you have to take your clothes to the Laundromat down the street. In the office equivalent, designers force employees to move around, too. They build in "functional inefficiencies"; they put kitchens and copiers and printers and libraries in places that can be reached only by a circuitous journey.

A more direct approach is to create an office so flexible that the kinds of people who need to spontaneously interact can actually be brought together. For example, the Ford Motor Company, along with a group of researchers from the University of Michigan, recently conducted a pilot project on the effectiveness of "war rooms" in software development. Previously, someone inside the company who needed a new piece of software written would have a series of meetings with the company's programmers, and the client and the programmers would send messages back and forth. In the war-room study, the company moved the client, the programmers, and a manager into a dedicated conference room, and made them stay there until the project was done. Using the war room cut the software-development time by two-thirds, in part because there was far less time wasted on formal meetings or calls outside the building: the people who ought to have been bumping into each other were now sitting next to each other.

Two years ago, the advertising agency TBWA\Chiat\Day moved into new offices in Los Angeles, out near the airport. In the preceding years, the firm had been engaged in a radical, and in some ways disastrous, experiment with a "nonterritorial" office: no one had a desk or any office equipment of his own. It was a scheme that courted failure by neglecting all the ways in which an office is a sort of neighborhood. By contrast, the new office is an almost perfect embodiment of Jacobsian principles of community. The agency is in a huge old warehouse, three stories high and the size of three football fields. It is informally known as Advertising City, and that's what it is: a kind of artfully constructed urban neighborhood. The floor is bisected by a central corridor called Main Street, and in the center of the room is an open space, with café tables and a stand of ficus trees, called Central Park. There's a basketball court, a game room, and a bar. Most of the employees are in snug workstations known as nests, and the nests are grouped together in neighborhoods that radiate from Main Street like Paris arrondissements. The top executives are situated in the middle of the room. The desk belonging to the chairman and creative director of the company looks out on Central Park. The offices of the chief financial officer and the media director abut the basketball court. Sprinkled throughout the building are meeting rooms and project areas and plenty of nooks where employees can closet themselves when they need to. A small part of the building is elevated above the main floor on a mezzanine, and if you stand there and watch the people wander about with their portable phones, and sit and chat in Central Park, and play basketball in the gym, and you feel on your shoulders the sun from the skylights and listen to the gentle buzz of human activity, it is quite possible to forget that you are looking at an office.

In "The Death and Life of Great American Cities," Jacobs wrote of the importance of what she called "public characters"—people who have the social position and skills to orchestrate the movement of information and the creation of bonds of trust:

A public character is anyone who is in frequent contact with a wide circle of people and who is sufficiently interested to make himself a public character. . . . The director of a settlement on New York's Lower East Side, as an example, makes a regular round of stores. He learns from the cleaner who does his suits about the presence of dope pushers in the neighborhood. He learns from the grocer that the Dragons are working up to something and need attention. He learns from the candy store that two girls are agitating the Sportsmen toward a rumble. One of his most important information spots is an unused breadbox on Rivington Street. . . . A message spoken there for any teen-ager within many blocks will reach his ears unerringly and surprisingly quickly, and the opposite flow along the grapevine similarly brings news quickly in to the breadbox.

A vital community, in Jacobs's view, required more than the appropriate physical environment. It also required a certain kind of person, who could bind together the varied elements of street life. Offices are no different. In fact, as office designers have attempted to create more vital workplaces, they have become increasingly interested in identifying and encouraging public characters.

One of the pioneers in this way of analyzing offices is Karen Stephenson, a business-school professor and anthropologist who runs a New York-based consulting company called Netform. Stephenson studies social networks. She goes into a company—her clients include J. P. Morgan, the Los Angeles Police Department, T.R.W., and I.B.M.—and distributes a questionnaire to its employees, asking about which people they have contact with. Whom do you like to spend time with? Whom do you talk to about new ideas? Where do you go to get expert advice? Every name in the company becomes a dot on a graph, and Stephenson draws lines between all those who have regular contact with each other. Stephenson likens her graphs to X-rays, and her role to that of a radiologist. What she's depicting is the firm's invisible inner mechanisms, the relationships and networks and patterns of trust that arise as people work together over time, and that are hidden beneath the organization chart. Once, for example, Stephenson was doing an "X-ray" of a Head Start organization. The agency was mostly female, and when Stephenson analyzed her networks she found that new hires and male staffers were profoundly isolated, communicating with the rest of the organization through only a handful of women. "I looked at tenure in the organization, office ties, demographic data. I couldn't see what tied the women together, and why the men were talking only to these women," Stephenson recalls. "Nor could the president of the organization. She gave me a couple of ideas. She said, `Sorry I can't figure it out.' Finally, she asked me to read the names again, and I could hear her stop, and she said, `My God, I know what it is. All those women are smokers.'" The X-ray revealed that the men—locked out of the formal power structure of the organization—were trying to gain access and influence by hanging out in the smoking area with some of the more senior women.

What Stephenson's X-rays do best, though, is tell you who the public characters are. In every network, there are always one or two people who have connections to many more people than anyone else. Stephenson calls them "hubs," and on her charts lines radiate out from them like spokes on a wheel. (Bernie the candy-store owner, on Jacobs's Hudson Street, was a hub.) A few people are also what Stephenson calls "gatekeepers": they control access to critical people, and link together a strategic few disparate groups. Finally, if you analyze the graphs there are always people who seem to have lots of indirect links to other people—who are part of all sorts of networks without necessarily being in the center of them. Stephenson calls those people "pulsetakers." (In Silicon Valleyspeak, the person in a sea of cubicles who pops his or her head up over the partition every time something interesting is going on is called a prairie dog: prairie dogs are pulsetakers.)

In the past year, Stephenson has embarked on a partnership with Steelcase, the world's largest manufacturer of office furniture, in order to use her techniques in the design of offices. Traditionally, office designers would tell a company what furniture should go where. Stephenson and her partners at Steelcase propose to tell a company what people should go where, too. At Steelcase, they call this "floor-casting."

One of the first projects for the group is the executive level at Steelcase's headquarters, a five-story building in Grand Rapids, Michigan. The executive level, on the fourth floor, is a large, open room filled with small workstations. (Jim Hackett, the head of the company, occupies what Steelcase calls a Personal Harbor, a black, freestanding metal module that may be—at seven feet by eight—the smallest office of a Fortune 500 C.E.O.) One afternoon recently, Stephenson pulled out a laptop and demonstrated how she had mapped the communication networks of the leadership group onto a seating chart of the fourth floor. The dots and swirls are strangely compelling—abstract representations of something real and immediate. One executive, close to Hackett, was inundated with lines from every direction. "He's a hub, a gatekeeper, and a pulsetaker across all sorts of different dimensions," Stephenson said. "What that tells you is that he is very strategic. If there is no succession planning around that person, you have got a huge risk to the knowledge base of this company. If he's in a plane accident, there goes your knowledge." She pointed to another part of the floor plan, with its own thick overlay of lines. "That's sales and marketing. They have a pocket of real innovation here. The guy who runs it is very good, very smart." But then she pointed to the lines connecting that department with other departments. "They're all coming into this one place," she said, and she showed how all the lines coming out of marketing converged on one senior executive. "There's very little path redundancy. In human systems, you need redundancy, you need communication across multiple paths." What concerned Stephenson wasn't just the lack of redundancy but the fact that, in her lingo, many of the paths were "unconfirmed": they went only one way. People in marketing were saying that they communicated with the senior management, but there weren't as many lines going in the other direction. The sales-and-marketing team, she explained, had somehow become isolated from senior management. They couldn't get their voices heard when it came to innovation—and that fact, she said, ought to be a big consideration when it comes time to redo the office. "If you ask the guy who heads sales and marketing who he wants to sit next to, he'll pick out all the people he trusts," she said. "But do you sit him with those people? No. What you want to do is put people who don't trust each other near each other. Not necessarily next to each other, because they get too close. But close enough so that when you pop your head up, you get to see people, they are in your path, and all of a sudden you build an inviting space where they can hang out, kitchens and things like that. Maybe they need to take a hub in an innovation network and place the person with a pulsetaker in an expert network—to get that knowledge indirectly communicated to a lot of people."

The work of translating Stephenson's insights onto a new floor plan is being done in a small conference room—a war room—on the second floor of Steelcase headquarters. The group consists of a few key people from different parts of the firm, such as human resources, design, technology, and space-planning research. The walls of the room are cluttered with diagrams and pictures and calculations and huge, blownup versions of Stephenson's X-rays. Team members stress that what they are doing is experimental. They don't know yet how directly they want to translate findings from the communications networks to office plans. After all, you don't want to have to redo the entire office every time someone leaves or joins the company. But it's clear that there are some very simple principles from the study of public characters that ought to drive the design process. "You want to place hubs at the center," Joyce Bromberg, the director of space planning, says. "These are the ones other people go to in order to get information. Give them an environment that allows access. But there are also going to be times that they need to have control—so give them a place where they can get away. Gatekeepers represent the fit between groups. They transmit ideas. They are brokers, so you might want to put them at the perimeter, and give them front porches"—areas adjoining the workspace where you might put little tables and chairs. "Maybe they could have swinging doors with white boards, to better transmit information. As for pulsetakers, they are the roamers. Rather than give them one fixed work location, you might give them a series of touchdown spots—where you want them to stop and talk. You want to enable their meandering."

One of the other team members was a tall, thoughtful man named Frank Graziano. He had a series of pencil drawings—with circles representing workstations of all the people whose minds, as he put it, he wanted to make "explicit." He said that he had done the plan the night before. "I think we can thread innovation through the floor," he went on, and with a pen drew a red line that wound its way through the maze of desks. It was his Hudson Street.

"The Death and Life of Great American Cities" was a controversial book, largely because there was always a whiff of paternalism in Jacobs's vision of what city life ought to be. Chelsea—the neighborhood directly to the north of her beloved West Village—had "mixtures and types of buildings and densities of dwelling units per acre... almost identical with those of Greenwich Village," she noted. But its long-predicted renaissance would never happen, she maintained, because of the "barriers of long, self-isolating blocks." She hated Chatham Village, a planned "garden city" development in Pittsburgh. It was a picturesque green enclave, but it suffered, in Jacobs's analysis, from a lack of sidewalk life. She wasn't concerned that some people might not want an active street life in their neighborhood; that what she saw as the "self-isolating blocks" of Chelsea others would see as a welcome respite from the bustle of the city, or that Chatham Village would appeal to some people precisely because one did not encounter on its sidewalks a "solid wave of conversation and animation." Jacobs felt that city dwellers belonged in environments like the West Village, whether they realized it or not.

The new workplace designers are making the same calculation, of course. The point of the new offices is to compel us to behave and socialize in ways that we otherwise would not—to overcome our initial inclination to be office suburbanites. But, in all the studies of the new workplaces, the reservations that employees have about a more social environment tend to diminish once they try it. Human behavior, after all, is shaped by context, but how it is shaped—and whether we'll be happy with the result—we can understand only with experience. Jane Jacobs knew the virtues of the West Village because she lived there. What she couldn't know was that her ideas about community would ultimately make more sense in the workplace. From time to time, social critics have bemoaned the falling rates of community participation in American life, but they have made the same mistake. The reason Americans are content to bowl alone (or, for that matter, not bowl at all) is that, increasingly, they receive all the social support they need—all the serendipitous interactions that serve to make them happy and productive—from nine to five.

Chapter 2 ■ Designs for Working

Questions for Discussion

1. One of the interesting things about Gladwell's essay is the way it gets you to look again at the spaces that surround you every day: from the city you live in to the school rooms you inhabit. How well are the spaces you live in every day designed for building trust, engagement, community, and productivity? What could be done to improve those spaces, along the lines that Gladwell discusses?

2. In adapting Jane Jacobs's theories about urban environments to the office place, managers have taken a paradigm from one field and applied it to another. How well does it guide the practice of office design? What parallels between the two does Gladwell draw?

3. Karen Stephenson's study (discussed near the end of the essay) is suggestive of the sort of work you might do yourself to test a hypothesis in the workplace. What other productivity-related issues can you record and measure in order to better understand your workplace or university?

4. If you have been taking this class for at least a week or two, it is likely that relationships and connections have developed in the classroom. And studies have shown that students tend to gravitate to specific areas of the classroom or even specific chairs, unless the teacher requires a special seating arrangement or takes some action to change students' chosen spots. Do you think there is a relationship between where you are sitting and with whom you have a connection—from mere acquaintances, to people you have spoken to outside of class, to people you would consider a friend? How much is the strength of that connection perhaps simply a factor of the physical distance between your chairs?

Try to replicate Stephenson's study. Distribute a copy of the class roster to everyone in the classroom and have them rate from 0 (no contact) to 5 (friendship) the level of connection they feel with others in the room. Then draw a seating chart on the board and compare some of the numbers that people put down with the physical space between them.

How well does Stephenson's study hold up for your class? What limited conclusions might you draw from your findings? How might those conclusions help to shape some action to change the dynamic of the classroom for the better, perhaps in order to promote more classroom participation?

A New Germ Theory

Judith Hooper

The dictates of evolution virtually demand that the causes of some of humanity's chronic and most baffling "noninfectious" illnesses will turn out to be pathogens—that is the radical view of a prominent evolutionary biologist.

A late-September heat wave enveloped Amherst College, and young people milled about in shorts or sleeveless summer frocks, or read books on the grass. Inside the red-brick buildings framing the leafy quadrangle students listened to lectures on Ellison and Emerson, on Paul Verlaine and the Holy Roman Empire. Few suspected that strains of the organism that causes cholera were growing nearby, in the Life Sciences Building. If they had known, they would probably not have grasped the implications. But these particular strains of cholera make Paul Ewald smile; they are strong evidence that he is on the right track. Knowing the rules of evolutionary biology, he believes, can change the course of infectious disease.

In a hallway of the Life Sciences Building an anonymous student has scrawled above a display of glossy photographs and vitae of the faculty, "We are the water; you are but the sponge." This is the home of Amherst's biology department, where Paul Ewald is a professor. He is also the author of the seminal book *Evolution of Infectious Disease* and of a long list of influential papers. Sandy-haired, trim, and handsome in an all-American way, he looks considerably younger than his forty-five years. Conspicuously outdoorsy for an academic, he would not seem out of place in an L. L. Bean catalogue, with a golden retriever by his side. Ewald rides his bike to the campus every day in decent weather—and in weather one might not consider decent—from the nearby hill village of Shutesbury, where he lives with his wife, Chris, and two teenage children in a restored eighteenth-century house.

As far as Ewald is concerned, Darwin's legacy is the most interesting thing on the planet. The appeal of evolutionary theory is that it is a grand unifying principle, linking all organisms, from protozoa to Presidents, and yet its essence is simple and transparent. "Darwin only had a couple of basic tenets," Ewald observed recently in his office. "You have heritable variation, and you've got differences in survival and reproduction among the variants. That's the beauty of it. It has to be true—it's like arithmetic. And if there is life on other planets, natural selection has to be the fundamental organizing principle there, too."

These Darwinian laws have led Ewald to a new theory: that diseases we have long ascribed to genetic or environmental factors—including some forms of heart disease, cancer, and mental illness—are in many cases actually caused by infections. Before we take up this theory, we need to spend a moment with Ewald's earlier work.

Ewald began in typical evolutionary terrain, studying hummingbirds and other creatures visible to the naked eye. It was on a 1977 field trip to study a species called Harris's sparrow in Kansas that a bad case of diarrhea laid him up for a few days and changed the course of his career. The more he

Used with permission of Atlantic Monthly Co, from *Atlantic Monthly*, February 1999 by Judith Hooper; permission conveyed through Copyright Clearance Center, Inc.

meditated on how Darwinian principles might apply to the organisms responsible for his distress—asking himself, for instance, what impact treating the diarrhea would have on the vast populations of bacteria evolving within his intestine—the more obsessed he became. Was his diarrhea a strategy used by the pathogen to spread itself, he wondered, or was it a defense employed by the host—his body—to flush out the invader? If he curbed the diarrhea with medication, would he be benefiting the invader or the host? Ewald's paper outlining his speculations about diarrhea was published in 1980, in the *Journal of Theoretical Biology*. By then Ewald was on his way to becoming the Darwin of the microworld.

"Ironically," he says, "natural selection was first recognized as operating in large organisms, and ignored in the very organisms in which it is especially powerful—the microorganisms that cause disease. The time scale is so much shorter and the selective pressures so much more intense. You can get evolutionary change in disease organisms in months or weeks. In something like zebras you'd have to wait many centuries to see it."

For decades medical science was dominated by the doctrine of "commensalism"—the notion that the pathogen-host relationship inevitably evolves toward peaceful coexistence, and the pathogen itself toward mildness, because it is in the germ's interest to keep its host alive. This sounds plausible, but it happens to be wrong. The Darwinian struggle of people and germs is not necessarily so benign. Evolutionary change in germs can go either way, as parasitologists and population geneticists have realized—toward mildness or toward virulence. It was Ewald's insight to realize what we might do about it.

Manipulating the Enemy

Say you're a disease organism—a rhinovirus, perhaps, the cause of one of the many varieties of the common cold; or the mycobacterium that causes tuberculosis; or perhaps the pathogen that immobilized Ewald with diarrhea. Your best bet is to multiply inside your host as fast as you can. However, if you produce too many copies of yourself, you'll risk killing or immobilizing your host before you can spread. If you're the average airborne respiratory virus, it's best if your host is well enough to go to work and sneeze on people in the subway.

Now imagine that host mobility is unnecessary for transmission. If you're a germ that can travel from person to person by way of a "vector," or carrier, such as a mosquito or a tsetse fly, you can afford to become very harmful. This is why, Ewald argues, insect-borne diseases such as yellow fever, malaria, and sleeping sickness get so ugly. Cholera uses another kind of vector for transmission: it is generally waterborne, traveling easily by way of fecal matter shed into the water supply. And it, too, is very ugly.

"Here's the [safety] hood where we handle the cholera," Jill Saunders explained as we toured the basement lab in Amherst's Life Sciences Building where cholera strains are stored in industrial refrigerators after their arrival from hospitals in Peru, Chile, and Guatemala. "We always wear gloves." A medical-school-bound senior from the Boston suburbs, Saunders is one of Ewald's honor students. As she guided me around, pointing out centrifuges, -80 degree freezers, and doors with BIOHAZARD warnings, we passed a closet-sized room as hot and steamy as the tropical zones where hemorrhagic fevers thrive. She said, "This is the incubation room, where we grow the cholera."

Cholera invaded Peru in 1991 and quickly spread throughout South and Central America, in the process providing a ready-made experiment for Ewald. On the day of my tour Saunders had presented to the assembled biology department her honors project, "Geographical Variations in the Virulence of *Vibrio cholerae* in Latin America." The data compressed in her tables and bar graphs were evidence for Ewald's central thesis: it is possible to influence a disease organism's evolution to your advantage. Saunders used a standard asSay, called ELISA, to measure the amount of toxin produced by different strains of cholera, thus inferring the virulence of *V. cholerae* variants from several Latin

American regions. Then she and Ewald looked at figures for water quality—what percentage of the population had potable water, for example—and looked for correlations. If virulent strains correlated with a contaminated water supply, and if, conversely, mild strains took over where the water was clean, the implication would be that *V. cholerae* becomes increasingly mild when it cannot use water as a vector. When the pathogen is denied easy access to new hosts through fecal matter in the water system, its transmission depends on infected people moving into contact with healthy ones. In this scenario the less-toxic variants would prevail, because these strains do not incapacitate or kill the host before they can be spread to others. If this turned out to be true, it would constitute the kind of evidence that Ewald expected to find.

The dots on Saunders's graphs made it plain that cholera strains are virulent in Guatemala, where the water is bad, and mild in Chile, where water quality is good. "The Chilean data show how quickly it can become mild in response to different selective pressures," Ewald explained. "Public-health people try to keep a disease from spreading in a population, and they don't realize that we can also change the organism itself. If you can make an organism very mild, it works like a natural vaccine against the virulent strains. That's the most preventive of preventive medicine: when you can change the organism so it doesn't make you sick." Strains of the cholera agent isolated from Texas and Louisiana produce such small amounts of toxin that almost no one who is infected with them will come down with cholera.

Joseph Schall, a professor of biology at the University of Vermont, offers a comment on Ewald's work: "If Paul is right, it may be that the application of an evolutionary theory to public health could save millions of lives. It's a stunning idea. If we're able to manipulate the evolutionary trajectory of our friends—domestic animals and crops—why not do the same with our enemies, with cholera, malaria, and HIV? As Thomas Huxley said when he read Darwin, "How stupid of me not to have thought of that before." I thought when I heard Paul's idea, "Gee, why didn't *I* think of that?"

Ewald put forward his virulence theories in *Evolution of Infectious Disease.* Today his book is on the syllabus for just about every college course in Darwinian medicine or its equivalent. "I regard him as a major figure in the field," says Robert Trivers, a prominent evolutionary biologist who holds professorships in anthropology and biology at Rutgers University. "It is a shame his work isn't better known to the public-health and medical establishments, who are willfully ignorant of evolutionary logic throughout their training." While praising Ewald's boldness and originality, some of his peers caution that his data need to be independently corroborated, and others object that his hypotheses are too crude to capture the teeming complexity of microbial evolution. "Evolutionary biologists have had very poor success in explaining how an organism evolves in response to its environment," says James Bull, an evolutionary geneticist at the University of Texas. "Trying to understand a two-species interaction should be even more complicated."

Recently, in any case, Ewald has adopted a new cause, far more radical but equally rooted in evolution. Let's call it Germ Theory, Part II. It offers a new way to think about the causes of some of humanity's chronic and most baffling illnesses. Ewald's view, to put it simply, is that the culprits will often turn out to be pathogens—that the dictates of evolution virtually demand that this be so.

The Case for Infection

Germ Theory, Part I, the edifice built by men like Louis Pasteur, Edward Jenner, and Robert Koch, took medicine out of the Dark Ages. It wasn't "bad air" or "bad blood" that caused diseases like malaria and yellow fever but pathogens transmitted by mosquitoes. Tuberculosis was famously tracked to an airborne pathogen, *Mycobacterium tuberculosis,* by Robert Koch, the great German scientist who in 1905 won a Nobel Prize for his work. Koch also revolutionized medical epidemiology by laying out his famous four postulates, which have set the standard for proof of infectivity up to the present day. The postulates dictate that a microbe must be (a) found in an animal (or person) with the disease;

(b) isolated and grown in culture; (c) injected into a healthy experimental animal, producing the disease in question; and then (d) recovered from the experimentally diseased animal and shown to be the same pathogen as the original.

By the early twentieth century the whole landscape had changed. Most of the common killer diseases, including smallpox, diphtheria, bubonic plague, flu, whooping cough, yellow fever, and TB, were understood to be caused by pathogens. Vaccines were devised against some, and by the 1950s antibiotics could easily cure many others. Smallpox was actually wiped off the face of the earth (if you don't count a few strains preserved in laboratories in the United States and Russia).

By the 1960s and 1970s the prevailing mood was one of optimism. Ewald is fond of quoting from a 1972 edition of a classic medical textbook: "The most likely forecast about the future of infectious disease is that it will be very dull." At least in the developed world, infectious diseases no longer seemed very threatening. Far scarier were the diseases that the medical world said were not infectious: heart disease, cancer, diabetes, and so on. No one foresaw the devastation of AIDS, or the serial outbreaks of deadly new infections such as Legionnaire's disease, Ebola and Marburg hemorrhagic fevers, antibiotic-resistant tuberculosis, "flesh-eating" staph infections, hepatitis C, and Rift Valley fever.

The infectious age is, we now know, far from over. Furthermore, it appears that many diseases we didn't think were infectious may be caused by infectious agents after all. Ewald observes,"By guiding researchers down one path, Koch's postulates directed them away from alternate ones. Researchers were guided away from diseases that might have been infectious but had little chance of fulfilling the postulates." That is, just because we couldn't readily discover their cause, we rather arbitrarily decided that the so-called chronic diseases of the late twentieth century must be hereditary or environmental or "multifactorial." And, Ewald contends, we have frequently been wrong.

Germ Theory, Part II, as conceived by Ewald and his collaborator, Gregory M. Cochran, flows from the timeless logic of evolutionary fitness. Coined by Darwin to refer to the fit between an organism and its environment, the term has come to mean the evolutionary success of an organism relative to competing organisms. Genetic traits that may be unfavorable to an organism's survival or reproduction do not persist in the gene pool for very long. Natural selection, by its very definition, weeds them out in short order. By this logic, any inherited disease or trait that has a serious impact on fitness must fade over time, because the genes that spell out that disease or trait will be passed on to fewer and fewer individuals in future generations. Therefore, in considering common illnesses with severe fitness costs, we may presume that they are unlikely to have a genetic cause. If we cannot track them to some hostile environmental element (including lifestyle), Ewald argues, then we must look elsewhere for the explanation. "When diseases have been present in human populations for many generations and still have a substantial negative impact on people's fitness," he says, "they are likely to have infectious causes."

Although its fitness-reducing dimensions are difficult to calculate, the ordinary stomach ulcer is the best recent example of a common ailment for which an infectious agent—to the surprise of almost everyone—turns out to be responsible.

When I visited him one afternoon, Ewald pulled off his shelves a standard medical textbook from the 1970s and opened the heavy volume to the entry on peptic ulcers. We squinted together at a gray field of small print punctuated by subheads in boldface. Under "Etiology" we scanned several pages: *environmental factors . . . smoking . . . diet . . . ulcers caused by drugs . . . aspirin . . . psychonomic factors . . . lesions caused by stress.* In the omniscient tone of medical texts, the authors concluded, "It is plausible to hypothesize a wealth of these factors. . . . " There was no mention of infection at all.

In 1981 Barry J. Marshall was training in internal medicine at the Royal Perth Hospital, in Western Australia, when he became interested in incidences of spiral bacteria in the stomach lining. The bacteria were assumed to be irrelevant to ulcer pathology, but Marshall and J. R. Warren, a histopathologist who had previously observed the bacteria, reviewed the records of patients whose stomachs were infected with large numbers of these bacteria. They noticed that when one patient was treated with tetracycline for unrelated reasons, his pain vanished, and an endoscopy revealed that his ulcer was gone.

An article by Marshall and Warren on their culturing of "unidentified curved bacilli" appeared in the British medical journal *The Lancet* in 1984, and was followed by other suggestive studies. For years, however, the medical establishment remained deaf to their findings, and around the world ulcer patients continued to dine on bland food, swear off stress, and swill Pepto-Bismol. Finally Marshall personally ingested a batch of the spiral bacteria and came down with painful gastritis, thereby fulfilling all of Koch's postulates.

There is now little doubt that *Helicobacter pylori,* found in the stomachs of a third of adults in the United States, causes inflammation of the stomach lining. In 20 percent of infected people it produces an ulcer. Nearly everyone with a duodenal ulcer is infected. *H. pylori* infections can be readily diagnosed with endoscopic biopsy tests, a blood test for antibodies, or a breath test. In 90 percent of cases the infections can be cured in less than a month with antibiotics. (Unfortunately, many doctors still haven't gotten the news. A Colorado survey found that 46 percent of patients seeking medical attention for ulcer symptoms are never tested for *H. pylori* by their physicians.)

Antibiotics Against Heart Disease?

Ewald closed the medical textbook on his knee. "This was published twenty years ago," he said. "If we looked up 'atherosclerosis' in a textbook from ten years ago, we'd find the same kind of things—stress, lifestyle, lots about diet, nothing about infection."

Heart disease is now being linked to *Chlamydia pneumoniae,* a newly discovered bacterium that causes pneumonia and bronchitis. The germ is a relative of *Chlamydia trachomatis,* which causes trachoma, a leading cause of blindness in parts of the Third World. *C. trachomatis* is perhaps more familiar to us as a sexually transmitted disease that, left untreated in women, can lead to scarring of the fallopian tubes, pelvic inflammatory disease, ectopic pregnancy, and tubal infertility.

Pekka Saikku and Maija Leinonen, a Finnish husband-and-wife team who have evoked comparisons to the Curies, discovered the new type of chlamydial infection in 1985, though its existence was not officially recognized until 1989. Saikku and Leinonen found that 68 percent of Finnish patients who had suffered heart attacks had high levels of antibodies to *C. pneumoniae,* as did 50 percent of patients with coronary heart disease, in contrast to 17 percent of the healthy controls. "We were mostly ignored or laughed at," Saikku recalls.

While examining coronary-artery tissues at autopsy in 1991, Allan Shor, a pathologist in Johannesburg, saw "pear-shaped bodies" that looked like nothing he'd ever seen before. He mentioned his observations to a microbiologist colleague, who had read about a new species of chlamydia with a peculiar pear shape. The colleague referred Shor to an expert on the subject, Cho-Chou Kuo, of the University of Washington School of Public Health, in Seattle. After Shor shipped Kuo the curious coronary tissue, Kuo found that the clogged coronary arteries were full of *C. pneumoniae.* Before long, others were reporting the presence of live *C. pneumoniae* in arterial plaque fresh from operating tables. Everywhere the bacterium lodges, it appears to precipitate the same grim sequence of events: a chronic inflammation, followed by a buildup of plaque that occludes the opening of the artery (or, in the case of venereal chlamydia, a buildup of scar tissue in the fallopian tube). Recently a team of pathologists at MCP-Hahnemann School of Medicine, in Philadelphia, found the same bacterium in the diseased sections of the autopsied brains of patients who had had late-onset Alzheimer's disease: it was present in seventeen of nineteen Alzheimer's patients and in only one of nineteen controls.

By the mid-1990s a radical new view was emerging of atherosclerosis as a chronic, lifelong arterial infection. "I am confident that this will reach the level of certainty of ulcer and *H. pylori,*" says Saikku, who estimates that at least 80 percent of all coronary heart disease is caused by the bacterium. Big questions remain, of course. Studies show that about 50 percent of U.S. adults carry antibodies to *C. pneumoniae*—but how many will develop heart disease? Even if heart patients can be shown to have antibodies to *C. pneumoniae,* and even if colonies of the bacteria are found living and breeding in diseased coronary arteries, is it certain that the germ *caused* the damage? Perhaps it's an innocent bystander, as some critics have proposed; or a secondary, opportunistic infection.

But suppose that a *Chlamydia pneumoniae* infection during childhood can initiate a silent, chronic infection of the coronary arteries, resulting in a "cardiovascular event" fifty years later. Could antibiotics help to address the problem?

A few early studies suggest they might. Researchers in Salt Lake City infected white rabbits with *C. pneumoniae,* fed them a modestly cholesterol-enhanced diet, killed them, and found thickening of the thoracic aortas, in contrast to the condition of uninfected controls fed the same diet. Additionally, treatment of infected rabbits with antibiotics in the weeks following infection prevented the thickening. Saikku and colleagues reported a similar finding, also in rabbits. Coronary patients in Europe who were treated with azithromycin not only showed a decline in antibodies and other markers of infection but in some studies had fewer subsequent cardiovascular events than patients who were given placebos. (These findings are preliminary; in a few years we may know more. The first major clinical trial is under way in the United States, sponsored by the National Institutes of Health and the Pfizer Corporation: 4,000 heart patients at twenty-seven clinical centers will be given either the antibiotic azithromycin or a placebo and followed for four years to gauge whether the antibiotic affects the incidence of further coronary events.)

Smoking, stress, cholesterol, and heredity all play a role in heart disease. But imagine if our No. 1 killer—with its vast culture of stress-reduction theories, low-fat diets, high-fiber cereals, cholesterol-lowering drugs, and high-tech bypass surgery—could in many instances be vanquished with an antibiotic. Numerous precedents exist for long-smoldering bacterial infections with consequences that appear months or years later. Lyme disease, leprosy, tuberculosis, and ulcers have a similar course. Ewald is confident that the association of *C. pneumoniae* and heart disease is real. He doesn't believe that the germ is an innocent bystander. "It reminds you a lot of gonorrhea in the 1890s," he says. "When they saw the organism there, people said, 'Well, we don't know if it's really causing the disease, or is just living there.' Every month the data are getting stronger. This is a smoking gun, just like *Helicobacter.*"

Evolutionary Byways

"I have a motto," Gregory Cochran told me recently. "'Big old diseases are infectious.' If it's common, higher than one in a thousand, I get suspicious. And if it's old, if it has been around for a while, I get suspicious."

The fact that Ewald has dared to conceive of a big theory for the medical sciences owes much to Cochran's contributions. A forty-five-year-old Ph.D. physicist who lives in Albuquerque with his wife and three small children, Cochran makes a living doing contract work on advanced optical systems for weaponry and other devices. Whereas Ewald is an academic insider, with department meetings to attend and honors theses to monitor, Cochran is a solo player, with an encyclopedic mind (he is a former College Bowl contestant) and a manner that verges on edginess. These days he spends a lot of time at his computer, as rapt as a conspiracy theorist, cruising Medline for new data on infectious diseases and, one imagines, almost cackling to himself when he finds something really good. Cochran's background in a field dominated by grand theories and universal laws may serve as a valuable counterpoint to the empirical and theory-hostile universe of the health sciences.

Ewald and Cochran encountered each other serendipitously, after Cochran decided to pursue a certain line of thinking about a very sensitive subject. "I was reading an article in *Scientific American* in 1992 about pathogens manipulating a host to get what they want," Cochran recalls. "It described a flowering plant infected by a fungus, and the fungus hijacks the plant's reproductive machinery so that instead of pollen it produces fungal spores. I thought, *Could it be?*" Cochran strayed from his field to try his hand at writing an article on biology—elaborating an audacious theory that human homosexuality might result from a "manipulation" of a host by a germ with its own agenda. He sent his draft to a prestigious biology journal, which sent it out to three scientists for peer review. Two were unconvinced, even appalled; the third was Paul Ewald, who thought the article was flawed but who was nonetheless impressed by the logic of the idea. The article was rejected, but Ewald and Cochran began their association.

To illustrate his thinking about infectiousness and disease, Cochran not long ago gave me a tour of his conceptual bins, into which he sorts afflictions according to their fitness impact. Remember that fitness can be defined as the evolutionary success of one organism relative to competing organisms. Only one thing counts: getting one's genes into the future. Any disease that kills host organisms before they can reproduce reduces fitness to zero. Obviously, fitness takes a major hit whenever the reproductive system itself is involved, as in the case of venereal chlamydia.

Consider a disease with a fitness cost of one percent—that is, a disease that takes a toll on survival or reproduction such that people who have it end up with one percent fewer offspring, on average, than the general population. That small amount adds up. If you have an inherited disease with a one percent fitness cost, in the next generation there will be 99 percent of the original number in the gene pool. Eventually the number of people with the disease will dwindle to close to zero—or, more precisely, to the rate produced by random genetic mutations: about one in 50,000 to one in 100,000.

We were considering the bin containing diseases that are profoundly antagonistic to fitness, with a fitness cost of somewhere between one and 10 percent by Cochran's calculations. My eye took in a catalogue of human ills—some familiar, some exotic, some historically fearsome but close to extinct, some lethal in the tropics but of little concern to inhabitants of the temperate zones. This list also showed prevailing medical opinion about cause. Each name of a disease was trailed by a lower-case letter: *i* (for infectious), *g* (genetic), *g+* (genetic defense against an infectious disease), *e* (caused by an environmental agent), or *u* (unknown). I read, "Atherosclerosis (*u*), . . . chlamydia (venereal) (*i*), cholera (*i*), diphtheria (*i*), endometriosis (*u*), filariasis (*i*), G6PD deficiency (*g+*), . . . hemoglobin E disease (*g+*), hepatitis B (*i*), hepatitis C (*i*), hookworm disease (*i*), kuru (*i*), . . . malaria (vivax) (*i*), . . . pertussis (*i*), pneumococcal pneumonia (*i*), polycystic ovary disease (*u*), scarlet fever (*i*), . . . tuberculosis (*i*), typhoid (*i*), yellow fever (*i*)."

Of the top forty fitness-antagonistic diseases on the list, thirty-three are known to be directly infectious and three are indirectly caused by infection; Cochran believes that the others will turn out to be infectious too. The most fitness-antagonistic diseases must be infectious, not genetic, Ewald and Cochran reason, because otherwise their frequency would have sunk to the level of random mutations. The exceptions would be either diseases that could be the effect of some new environmental factor (radiation or smoking, for example), or genetic diseases that balance their fitness cost with a benefit. Sickle-cell anemia is one example of the latter.

Though sickle-cell anemia is strictly heritable according to Mendelian laws, it is widely believed to have persisted in the population in response to infectious selective pressures. It heads the list of genetic diseases that Ewald dubs "self-destructive defenses," in which a disease fatal in its homozygous form (two copies of the gene) carries an evolutionary advantage to heterozygous carriers (with one copy), protecting against a terrible infection: in this case falciparum malaria, common in Africa. Similarly, cystic fibrosis, some argue, evolved in northern Europe as a defense against *Salmonella typhi,* the cause of typhoid fever. Infection thus explains why these deadly genetic diseases have remained in the human gene pool when they should have died out.

But what about something like atherosclerosis? I asked. Leaving aside the evidence concerning *C. pneumoniae,* it is not apparent why a genetic cause for atherosclerosis should be dismissed out of hand on evolutionary grounds. If it hits people in midlife or later, after they have launched their genes, how could it possibly affect fitness?

Cochran's response illustrates some of the intricacies of evolutionary thinking. "Well, obviously, it's not as bad as a disease that kills you before puberty, but I think it does have a fitness cost. First of all, it's *really* common. Second, people think that all you have to do to pass your genes along is have children, but that's not true. You still need to raise the offspring to adulthood. In a hunter-gatherer or subsistence-farming culture, the fitness impact of dying in midlife might be considerable, especially during bad times, like famines. You've got to feed your family. Also, cardiovascular disease is a leading cause of impotence, and any disease that makes males impotent at age forty-five has got to affect reproduction somewhat."

But Fifty-Year-Olds? Sixty-Year-Olds?

Grandmothers do a large proportion of the food-gathering in some tribal cultures, according to recent anthropological reports. "They aren't hampered by babies anymore, and they don't have to go around chucking spears like the men," says George C. Williams, a professor emeritus of ecology and evolution at the State University of New York at Stonybrook, and one of the pillars of modern evolutionary biology. "They contribute substantially to the family diet." If long-lived elders historically have made a difference by fostering the survival of their descendants, and therefore their genes, Cochran figures, then a disease that kills sixty-year-olds could have a fitness impact of around one percent.

The Cause-and-Effect Conundrum

"Know what that is?" Ewald asked. We were standing in the main corridor of the Life Sciences Building, gazing up at a decorative metalwork frieze that runs along the walls just above door height. A pair of hummingbirds chase each other in a circle. A human eye and an octopus eye face off. A human hand is juxtaposed with a chimpanzee hand. Ewald pointed to something that looked like a daddy longlegs with a video camera for a head. "Some kind of insect?" I ventured. "It's a virus," he said. "See, it's like a spaceship. That"—he pointed at the head—"is its DNA. It injects it inside the cell."

There is something unsettling and fascinating about a virus, an organism that is neither strictly alive nor strictly inanimate, and that replicates by sneaking inside a host cell and commandeering its machinery. "Viruses are essentially bits of nucleic acid—either DNA or RNA—wrapped in a protein capsule," Ewald explained. "A retrovirus, like HIV, is an RNA virus with a protein called reverse transcriptase built into it, and once it gets into a cell, it uses the reverse transcriptase to make a DNA copy of its RNA. This viral DNA copy can insert itself into our DNA, where it can be read by our protein-making machinery the same way our own instructions are read."

The modus operandi of the world's most feared virus, HIV, is clever, killing its hosts very, very slowly. A sexually transmitted pathogen, without the luxury of being spread through sneezes or coughs, must await its few opportunities patiently; if those infected have no symptoms and don't know they are sick, so much the better. A mild, chronic form of AIDS had in all likelihood been around for centuries in Africa, according to Ewald. Suddenly in the 1970s—owing to changing patterns of sexual activity and to population movements—deadly strains spread in the population of Central and East Africa.

HIV has an extremely high mutation rate, which means that it is continually evolving, even within a single patient, producing competing strains that fight for survival against the weapons produced by the immune system. If selective pressures—in this case a high potential sexual transmission—have forced the virus to evolve toward virulence, the opposite selective pressures could do the reverse. Conceivably, we could "tame" HIV, encouraging it to evolve toward comparative harmlessness. It was already known that preventive measures such as safe sex, fewer partners, clean needles, and so forth could curb the spread of the disease. But Ewald pointed out early on that social modification was a far more potent weapon than anyone realized. Once HIV was cut off from easy access to new hosts, milder strains would flourish—ones that the host could tolerate for longer and longer periods.

Indeed, Ewald argues, given limited public-health budgets, it might make sense to put more money into transmission-prevention programs and less into the search for vaccines. (He also has strong opinions about how drugs should be used to treat AIDS. He asserts that every time we use an antiviral drug like AZT, we produce an array of AZT-resistant HIVs in the population; if viral evolution is taken into account, antiviral drugs can be used more judiciously.)

Ewald's theories tilt him decidedly toward the optimistic camp. Even in the absence of a vaccine the AIDS epidemic will not inevitably worsen; it can be curbed without reducing transmission to zero. A natural experiment now occurring in Japan, he says, could be a test case for his theories. In the early 1990s highly virulent strains of HIV from Thailand took root in Japan, but Ewald predicts that low rates of sexual transmission in that country—due to widespread condom use and other factors—will act as a selective pressure on these strains so that they evolve toward mildness. If this is true, the trend should become evident over the next ten years.

Like HIV, many other viruses have an indolent course, with a long latency between infection and the development of symptoms. Herpes zoster, the agent of chicken pox, lingers in the body forever, capable of erupting as painful shingles decades later. There are also so-called hit-and-run infections, in which a pathogen or its products disrupt the body's immunological surveillance system; once the microbes are gone (or when they are present in such low frequency as to be undetectable), the immune response stays stuck in the "on" mode, causing a lingering inflammation. By the time symptoms occur, the microorganism itself has disappeared, and its genome will not be detectable in any tissue.

"The health sciences are still grappling with the masking effects of long delays between the onset of infection and the onset of disease," Ewald says. "Any time you have hit-and-run infections, slow viruses, lingering or relapsing infections, or a time lag between infection and symptoms, the cause and effect is going to be very cryptic. You won't find these newly recognized infections by the methods we used to find old infectious diseases. We have to be ready to think of all sorts of new, clever ways to identify pathogens. We will have to abandon Koch's postulates in some cases."

The Great Synthesizer

As of this writing, the ideas at the core of Germ Theory, Part II, have been presented by Ewald mostly in the form of lectures, and in communications with colleagues. The papers in which the ideas will be formally articulated are in preparation. Given Ewald's prominence, the ideas are bound to cause a stir. They will also draw criticism. In the medical sciences, where "theory" is a bad word and "Stick to the data" is the reigning motto, Ewald will come under particular scrutiny because his hypothesis arrives detached from a vast corpus of laboratory data. It is helpful to think of Ewald as continuing the tradition of the great scientific synthesizers. Darwin himself was a synthesizer extraordinaire, who composed the thesis of *The Origin of Species* largely out of hundreds of odds and ends contributed by others, from pigeon breeders to naturalists. "Professor So-and-so has observed . . . " is a recurring motif in Darwin's book.

Ewald's theory about evolution and infectiousness provides a framework that potentially unites diverse research on the front lines of various afflictions. Ulcers and heart disease have already been mentioned. Here are two more: cancer and mental illness.

In 1910 a man named Peyton Rous discovered the eponymous Rous sarcoma virus, demonstrating that chickens infected with it developed cancer. Over the years many other cancer viruses have been discovered in animals. And yet until 1979, despite broad hints from the animal world, not a single human cancer was generally accepted as infectious. Rous had been lucky: his chickens became sick only two weeks after infection. Human cancers follow a more languorous course, which means that by the time symptoms show up, any infectious causation may well be buried under a lifetime of irrelevant risk factors.

In 1979 HTLV-1, a retrovirus endemic in parts of Asia, Africa, and the Caribbean, and transmitted either sexually or from mother to child, was linked to certain leukemias and lymphomas; the cancer appeared decades after infection. The Epstein-Barr virus (the agent that causes mononucleosis) has now been associated with some B-cell lymphomas, with a nasopharyngeal cancer common in south China, and with Burkitt's lymphoma, a deadly childhood cancer of Africa. Some 82 percent of all cases of cervical cancer have been associated with the sexually transmitted human *papilloma* virus, a once relatively innocent-seeming pathogen responsible for genital warts.

H. pylori, the ulcer pathogen, confers a sixfold greater risk of stomach cancer, and accounts for at least half of all stomach cancers. Also, the lymphoid tissue of the stomach can produce a low-grade gastric lymphoma under the influence of this bacterium. Early reports indicate that the lymphoma is cured in 50 percent of cases by resolving the *H. pylori* infection—which may mark the first time in medical history that cancer has been cured with an antibiotic.

Hepatitis B and C, two of the ever-growing alphabet soup of hepatic diseases, have been linked to liver cancer. Herpes virus 8 has recently been discovered to be the cause of Kaposi's sarcoma. "There is no reason to believe that this flurry of discovery has now completed the list of infectious agents of cancer," Ewald says.

Among the many known animal cancer viruses is a closely studied retrovirus known as mouse mammary tumor virus (MMTV), which causes mammary-gland cancer in mice. This virus is transmitted from mother to offspring through mother's milk, lying latent in the daughter's mammary tissue until activated by hormones during her own lactation. Could such a virus be a factor in human breast cancer? In the mid-1980s researchers announced that they had found in malignant human breast tumors a DNA sequence resembling MMTV, but the excitement waned when the same sequence was found in normal breast tissue as well. Interest has been revived by the research of Beatriz G-T. Pogo, a professor in the departments of medicine and microbiology at Mount Sinai School of Medicine, in New York. Examining some 400 to 500 breast-cancer samples, she has found DNA sequences resembling MMTV that are not present in normal tissue or in other human cancers. She remains guarded about the implications.

Can You "Catch" Schizophrenia?

Microbes obviously can cause mental disorders—as syphilitic dementia, to name but one example, makes brutally clear. But most post-Freudian discussions of psychiatric dysfunction have tended not to invoke infection. Recently, however, some cases of childhood obsessive compulsive disorder (OCD) have hinted at a new set of possibilities. Children who have this disease may compulsively count the crayons in their book bags over and over again, or meticulously avoid each crack in the pavement, in order to ward off some imagined evil. Susan E. Swedo, of the National Institute of Mental Health, in Bethesda, Maryland, noticed strong resemblances between OCD and a disease called Sydenham's chorea, formerly known as Saint Vitus's dance, which, like rheumatic heart disease, is a rare complication of an untreated streptococcal infection. Streptococcal antibodies find their way into the brain and attack a region called the basal ganglia, causing characteristic clumsiness and arm-flapping movements along with obsessions, compulsions, senseless rituals, and *idées fixes*. Could some cases of childhood OCD be a milder version of this illness? The hunch paid off. In the early 1990s a new syndrome, known as PANDAS (pediatric autoimmune neuropsychiatric disorders associated with streptococcus), was recognized.

Some children with OCD get better when they are given intravenous immunoglobulin or undergo therapeutic plasma exchange to remove the antibodies from their blood. It is not known whether adult-onset OCD—whose most famous avatar was the germ-phobic Howard Hughes—also results from some sort of infection. But it is certainly provocative that a mental disorder can result from a lingering immune response. The phenomenon makes some people wonder about schizophrenia.

For years, amid the smorgasbord of theories about the etiology of schizophrenia, there has been recurring speculation about a schizophrenia virus. Karl Menninger wondered in the 1920s if schizophrenia might result from a flu infection. Later researchers pointed to data that showed seasonal and geographic patterns in the births of schizophrenics, suggestive of infection—though it must be said that the viral theorists were largely regarded as inhabiting the fringe. Genetic theories grabbed center stage, and by the 1990s most researchers were pinning their hopes on the genetic markers being identified in the Human Genome Project.

In Ewald and Cochran's view, evolutionary laws dictate that infection must be a factor in schizophrenia. "They announced they had the gene for schizophrenia, and then it turned out not to be true," Cochran said one day when I mentioned genetic markers. "I think they found and unfound the gene for depression about six times. Nobody's found a gene yet for any common mental illness. Maybe instead of the Human Genome Project we should have the Human Germ Project." Cochran is endorsing a suggestion made by several scientists in a recent issue of *Nature*. "I don't mean to say that the Human Genome Project isn't worthwhile for many reasons, but all the genes we've found have been for *rare* diseases. I don't think the common diseases are going to turn out that way."

Schizophrenia affects about one percent of the population, and thus in Ewald and Cochran's scheme is too common for a genetic disease that profoundly impairs fitness. As noted, the background mutation rate—the rate at which a gene spontaneously mutates—is typically about one in 50,000 to one in 100,000. Not surprisingly, genetic diseases that are severely fitness-impairing (for example, achondroplastic dwarfism) tend to have roughly the same odds, depending on the gene. (In a few cases, however, the gene involved may be especially error-prone, resulting in a higher frequency of mishaps. One of the most common genetic diseases, Duchenne's muscular dystrophy, afflicts boys at a rate of one in 7,000, reflecting the fragility of an uncommonly long gene.)

From the fitness perspective, schizophrenia is a catastrophe. It is estimated that male schizophrenics have roughly half as many offspring as the general population has. Female schizophrenics have roughly 75 percent as many. Schizophrenia should therefore approach the level of a random mutation after many generations. (To explain this away, some genetic theorists have proposed that in hunter-gatherer cultures schizophrenics were the tribal shamans—desirable as sexual partners—and thus did not incur a reproductive disadvantage.)

No one has found a schizophrenia virus yet, but some think they may be close. Following a tip from Ewald and Cochran, I typed "Borna virus" into my online search engine and ended up with a stack of scientific papers. Borna virus was first recognized as the cause of a neurologic disease in horses, and can infect nearly all warm-blooded animals, from birds to primates. Horses and other animals infected with Borna virus may exhibit depressed or apathetic behavior, weakness of the legs, abnormal body postures, or a staggering gait. Borna-infected laboratory rats exhibit learning disorders, exaggerated startle responses, and hyperactivity, among other things.

Royce Waltrip, an associate professor of psychiatry at the University of Mississippi with an expertise in virology, studies Borna virus. Despite being leery of a rash of inconsistent studies associating Borna virus with schizophrenia, Waltrip believes that "there is something there, though I don't know if it's a perinatal infection or an adult infection or what." When he started looking for antibodies to Borna in mental patients, he found that 14 percent of the schizophrenic patients had antibodies to two or three Borna proteins, whereas none of the healthy controls did. Waltrip speculates that Borna virus is not *the* cause of schizophrenia. "I think that schizophrenia is an etiologically heterogeneous disease," he said. "I think there are a finite number of ways the brain can respond to injury. There are probably different routes to schizophrenia, and there is probably more than one infectious pathway." One route, he hypothesizes, is Borna virus.

Ewald and Cochran do not doubt that multiple pathogens or multiple factors may be implicated in some broad disease syndromes, among them schizophrenia. But they worry, in general, that the "multifactorial" argument has become too facile a response. "That's what they *always* say when they don't know the cause of a disease," Cochran said on the phone. "They say it's *multifactorial.* Ulcers and heart disease were supposed to be multifactorial. But they're infections! Tuberculosis was supposed to be multifactorial. It's an infection!"

I happened to be visiting Ewald in his office when Cochran called, so we were having a three-way conversation, with Cochran's voice echoing over the static on a speaker phone. Outside the window the scene was shifting subtly into mid-autumn. Patches of orange and rust speckled the blue-green flanks of the Holyoke hills, and the students on the playing fields were wearing sweatpants.

But what about random accidents in utero as a cause of schizophrenia? I asked. Some kind of damage to the wiring?

"You'd have to say what caused the damage," Ewald responded, pointing out that the word "random" is often used to refer to something we haven't been able to understand. He noted once again how widespread schizophrenia is. "At this frequency—one percent of the population—we'd expect that natural selection would have led to protective mechanisms."

The same holds true for severe depression, Ewald believes. A tendency toward suicide doesn't make evolutionary sense in a world of organisms driven by the twin urgencies of survival and reproduction. The relentless engine of natural selection should have eliminated any genes that infringed on them. So why are these fitness-antagonistic traits still around?

This leads to a subject that Ewald is not shy about bringing up in discussions with colleagues and in professional lectures: homosexuality. Various pieces of evidence have been adduced in recent years, by prominent researchers, for some sort of genetic component to homosexuality. The question arises as to whether natural selection would sustain a homosexual trait in the gene pool for any length of time. The best estimates of the fitness cost of homosexuality hover around 80 percent: in other words, gay men (in modern times, at least) have only 20 percent as many offspring as heterosexuals have. Simple math shows how quickly an evolutionarily disadvantageous trait like this should dwindle, if it is a simple genetic phenomenon. The researchers Richard Pillard, at the Boston University School of Medicine, and Dean Hamer, at the National Cancer Institute, are not persuaded that natural selection would necessarily have eliminated a homosexual trait, and offer ingenious counterarguments. (And they note that historically the fitness cost may not have been very high, when gay men stayed in the closet, married, and had children.)

No one, of course, has ever isolated a bacterium or a virus responsible for sexual orientation, and speculations about the manner in which such an agent would be transmitted can be nothing more than that. But Ewald and Cochran contend that the severe "fitness hit" of homosexuality is a red flag that should not be ignored, and that an infectious process should at least be explored. "It's a very sensitive subject,"Ewald admits, "and I don't want to be accused of gay-bashing. But I think the idea is viable. What scientists are supposed to do is evaluate an idea on the soundness of the logic and the testing of the predictions it can generate."

The Search for Telltale Signs

After I had spent time talking to Ewald and Cochran and reading back issues of the journal *Emerging Infectious Diseases*, everything began to look infectious to me. The catalogue of suspected chronic diseases caused by infection, according to David A. Relman, an assistant professor of medicine, microbiology, and immunology at Stanford University, now includes "sarcoidosis, various forms of inflammatory bowel disease, rheumatoid arthritis, systemic lupus erythematosus, Wegener's granulomatosis, diabetes mellitus, primary biliary cirrhosis, tropical sprue, and Kawasaki disease." Ewald and Cochran's list of likely suspects would include all of the above plus many forms of heart disease, arteriosclerosis, Alzheimer's disease, many if not most forms of cancer, multiple sclerosis, most major psychiatric diseases, Hashimoto's thyroiditis, cerebral palsy, polycystic ovary disease, and perhaps obesity and certain eating disorders. From an evolutionary perspective, Cochran says, anorexia is strikingly inimical to the survival principle. "I mean, *not to eat*—what would cause that?"

"In all these situations you look for little signs of infectious spread," Ewald said in his office. "Is there geographic variation? Temporal variation? Does it go up or down across decades? Multiple sclerosis seems pretty clearly infectious, because you have these island populations where there was no MS and then you see it spread like a wave through the population. And you have this latitudinal gradient . . . "

"Yes!" Cochran burst from the speaker phone. "The farther you get from the Equator, the more common it is. It's three to four times more common if you grow up in Ontario than if you grow up in Mississippi. Some people have tried to say that's because Canadians are genetically different from Americans."

I downloaded a paper about extremely high rates of multiple sclerosis in the Shetland and Orkney Islands and other regions of Scotland, and I made a mental note of the many Canadian Web sites devoted to MS. Like other autoimmune diseases, MS looks suspiciously infectious for a number of reasons: epidemiological evidence of childhood exposure to disease agents, geographic clusters, abnormal immune responses to a variety of viruses, resemblances to animal models and human diseases with a relapsing-remitting course. And, in fact, a virus has been nominated: the human herpes virus 6, the agent of roseola infantum, a very mild disease of childhood. The connection, however, is by no means proved.

"No doubt everywhere people look there will be more and more examples of chronic diseases with infectious etiology," says Stephen S. Morse, an expert in infectious diseases at the Columbia University School of Public Health. "*Helicobacter* is probably the tip of the iceberg." Although we have wielded the tools of microbial cultivation for a hundred years, much of the microbial world is still as mysterious as an alien planet. "It has been estimated that only 0.4 percent of all extant bacterial species have been identified," David Relman has written. "Does this remarkable lack of knowledge pertain to the subset of microorganisms both capable of and accomplished in causing human disease?" Even the germs that inhabit our bodies—the so-called "human commensal flora," such as the swarming populations of organisms that live in the spaces between our teeth—are largely unknown, he points out. Most of them are presumably benign, up to a point. There are disquieting suggestions in the literature of a link between bacteria in dental plaque and coronary disease.

"Some people think it's scary to have these time bombs in our bodies," Ewald says, "but it's also encouraging—because if it's a disease organism, then there's probably something we can do about it. The textbooks say, In 1900 most people died of infectious diseases, and today most people don't die of infectious disease; they die of cancer and heart disease and Alzheimer's and all these things. Well, in ten years I think the textbooks will have to be rewritten to say, "Throughout history most people have died of infectious disease, and most people continue to die of infectious disease.""

Chapter 2 ■ A New Germ Theory

Questions for Discussion

1. How has the germ theory led to changes in the practice of treating common illnesses, such as ulcer? What does this suggest about the way in which paradigms guide our plans of action?

2. As a paradigm, how well does Ewald's "New Germ Theory" support innovative research? Where does it seem to lead down a blind alley? Where does the paradigm liberate our thinking and where might it over-determine it?

3. Can you think of any potential candidates of diseases that might be explained by the germ theory? How would you try to investigate this new hypothesis?

Breaking the Global Warming Gridlock

Daniel Sarewitz and Roger Pielke Jr.

*B*oth sides on the issue of greenhouse gases frame their arguments in terms of science, but each new scientific finding only raises new questions—dooming the debate to be a pointless spiral. It's time, the authors argue, for a radically new approach: if we took practical steps to reduce our vulnerability to today's weather, we would go a long way toward solving the problem of tomorrow's climate.

In the last week of October, 1998, Hurricane Mitch stalled over Central America, dumping between three and six feet of rain within forty-eight hours, killing more than 10,000 people in landslides and floods, triggering a cholera epidemic, and virtually wiping out the economies of Honduras and Nicaragua. Several days later some 1,500 delegates, accompanied by thousands of advocates and media representatives, met in Buenos Aires at the fourth Conference of the Parties to the United Nations Framework Convention on Climate Change. Many at the conference pointed to Hurricane Mitch as a harbinger of the catastrophes that await us if we do not act immediately to reduce emissions of carbon dioxide and other so-called greenhouse gases. The delegates passed a resolution of "solidarity with Central America" in which they expressed concern "that global warming may be contributing to the worsening of weather" and urged "governments, . . . and society in general, to continue their efforts to find permanent solutions to the factors which cause or may cause climate events." Children wandering bereft in the streets of Tegucigalpa became unwitting symbols of global warming.

But if Hurricane Mitch was a public-relations gift to environmentalists, it was also a stark demonstration of the failure of our current approach to protecting the environment. Disasters like Mitch are a present and historical reality, and they will become more common and more deadly regardless of global warming. Underlying the havoc in Central America were poverty, poor land-use practices, a degraded local environment, and inadequate emergency preparedness—conditions that will not be alleviated by reducing greenhouse-gas emissions.

At the heart of this dispiriting state of affairs is a vitriolic debate between those who advocate action to reduce global warming and those who oppose it. The controversy is informed by strong scientific evidence that the earth's surface has warmed over the past century. But the controversy, and the science, focus on the wrong issues, and distract attention from what needs to be done. The enormous scientific, political, and financial resources now aimed at the problem of global warming create the perfect conditions for international and domestic political gridlock, but they can have little effect on the root causes of global environmental degradation, or on the human suffering that so often accompanies it. Our goal is to move beyond the gridlock and stake out some common ground for political dialogue and effective action.

Framing the Issue

In politics everything depends on how an issue is framed: the terms of debate, the allocation of power and resources, the potential courses of action. The issue of global warming has been framed by a single question: Does the carbon dioxide emitted by industrialized societies threaten the earth's climate? On one side are the doomsayers, who foretell environmental disaster unless carbon-dioxide emissions are immediately reduced. On the other side are the cornucopians, who blindly insist that society can continue to pump billions of tons of greenhouse gases into the atmosphere with no ill effect, and that any effort to reduce emissions will stall the engines of industrialism that protect us from a Hobbesian wilderness. From our perspective, each group is operating within a frame that has little to do with the practical problem of how to protect the global environment in a world of six billion people (and counting). To understand why global-warming policy is a comprehensive and dangerous failure, therefore, we must begin with a look at how the issue came to be framed in this way. Two converging trends are implicated: the evolution of scientific research on the earth's climate, and the maturation of the modern environmental movement.

Since the beginning of the Industrial Revolution the combustion of fossil fuels—coal, oil, natural gas—has powered economic growth and also emitted great quantities of carbon dioxide and other greenhouse gases. More than a century ago the Swedish chemist Svante Arrhenius and the American geologist T. C. Chamberlin independently recognized that industrialization could lead to rising levels of carbon dioxide in the atmosphere, which might in turn raise the atmosphere's temperature by trapping solar radiation that would otherwise be reflected back into space—a "greenhouse effect" gone out of control. In the late 1950s the geophysicist Roger Revelle, arguing that the world was making itself the subject of a giant "geophysical experiment," worked to establish permanent stations for monitoring carbon-dioxide levels in the atmosphere. Monitoring documented what theory had predicted: atmospheric carbon dioxide was increasing.

In the United States the first high-level government mention of global warming was buried deep within a 1965 White House report on the nation's environmental problems. Throughout the 1960s and 1970s global warming—at that time typically referred to as "inadvertent modification of the atmosphere," and today embraced by the term "climate change"—remained an intriguing hypothesis that caught the attention of a few scientists but generated little concern among the public or environmentalists. Indeed, some climate researchers saw evidence for global cooling and a future ice age. In any case, the threat of nuclear war was sufficiently urgent, plausible, and horrific to crowd global warming off the catastrophe agenda.

Continued research, however, fortified the theory that fossil-fuel combustion could contribute to global warming. In 1977 the nonpartisan National Academy of Sciences issued a study called *Energy and Climate,* which carefully suggested that the possibility of global warming "should lead neither to panic nor to complacency." Rather, the study continued, it should "engender a lively sense of urgency in getting on with the work of illuminating the issues that have been identified and resolving the scientific uncertainties that remain." As is typical with National Academy studies, the primary recommendation was for more research.

In the early 1980s the carbon-dioxide problem received its first sustained attention in Congress, in the form of hearings organized by Representative Al Gore, who had become concerned about global warming when he took a college course with Roger Revelle, twelve years earlier. In 1983 the Environmental Protection Agency released a report detailing some of the possible threats posed by the anthropogenic, or human-caused, emission of carbon dioxide, but the Reagan Administration decisively downplayed the document. Two years later a prestigious international scientific conference in Villach, Austria, concluded that climate change deserved the attention of policymakers worldwide. The following year, at a Senate fact-finding hearing stimulated by the conference, Robert Watson, a climate scientist at NASA, testified, "Global warming is inevitable. It is only a question of the magnitude and the timing."

At that point global warming was only beginning to insinuate itself into the public consciousness. The defining event came in June of 1988, when another NASA climate scientist, James Hansen, told Congress with "ninety-nine percent confidence" that "the greenhouse effect has been detected, and it is changing our climate now." Hansen's proclamation made the front pages of major newspapers, ignited a firestorm of public debate, and elevated the carbon-dioxide problem to pre-eminence on the environmental agenda, where it remains to this day. Nothing had so galvanized the environmental community since the original Earth Day, eighteen years before.

Historically, the conservation and environmental movements have been rooted in values that celebrate the intrinsic worth of unspoiled landscape and propagate the idea that the human spirit is sustained through communion with nature. More than fifty years ago Aldo Leopold, perhaps the most important environmental voice of the twentieth century, wrote, "We face the question whether a still higher 'standard of living' is worth its cost in things natural, wild, and free. For us of the minority, . . . the chance to find a pasque-flower is a right as inalienable as free speech." But when global warming appeared, environmentalists thought they had found a justification better than inalienable rights—they had found facts and rationality, and they fell head over heels in love with science.

Of course, modern environmentalists were already in the habit of calling on science to help advance their agenda. In 1967, for example, the Environmental Defense Fund was founded with the aim of using science to support environmental protection through litigation. But global warming was, and is, different. It exists as an environmental issue only because of science. People can't directly sense global warming, the way they can see a clear-cut forest or feel the sting of urban smog in their throats. It is not a discrete event, like an oil spill or a nuclear accident. Global warming is so abstract that scientists argue over how they would know if they actually observed it. Scientists go to great lengths to measure and derive something called the "global average temperature" at the earth's surface, and the total rise in this temperature over the past century—an increase of about six tenths of a degree Celsius as of 1998—does suggest warming. But people and ecosystems experience local and regional temperatures, not the global average. Furthermore, most of the possible effects of global warming are not apparent in the present; rather, scientists predict that they will occur decades or even centuries hence. Nor is it likely that scientists will ever be able to attribute any isolated event—a hurricane, a heat wave—to global warming.

A central tenet of environmentalism is that less human interference in nature is better than more. The imagination of the environmental community was ignited not by the observation that greenhouse-gas concentrations were increasing but by the scientific conclusion that the increase was caused by human beings. The Environmental Defense Fund, perhaps because of its explicitly scientific bent, was one of the first advocacy groups to make this connection. As early as 1984 its senior scientist, Michael Oppenheimer, wrote on the op-ed page of *The New York Times,*

> With unusual unanimity, scientists testified at a recent Senate hearing that using the atmosphere as a garbage dump is about to catch up with us on a global scale. . . . Carbon dioxide emissions from fossil fuel combustion and other "greenhouse" gases are throwing a blanket over the Earth. . . . The sea level will rise as land ice melts and the ocean expands. Beaches will erode while wetlands will largely disappear. . . . Imagine life in a sweltering, smoggy New York without Long Island's beaches and you have glimpsed the world left to future generations.

Preserving tropical jungles and wetlands, protecting air and water quality, slowing global population growth—goals that had all been justified for independent reasons, often by independent organizations—could now be linked to a single fact, anthropogenic carbon-dioxide emissions, and advanced along a single political front, the effort to reduce those emissions. Protecting forests, for example, could help fight global warming because forests act as "sinks" that absorb carbon dioxide. Air pollution could be addressed in part by promoting the same clean-energy sources that would reduce carbon-dioxide emissions. Population growth needed to be controlled in order to reduce demand for fossil-fuel combustion. And the environmental community could reinvigorate its energy-conservation agenda, which had flagged since the early 1980s, when the effects of the second Arab oil shock wore off. Senator Timothy Wirth, of Colorado, spelled out the strategy in 1988: "What we've got to do in energy conservation is try to ride the global warming issue. Even if the theory of global warming is wrong, to have approached global warming as if it is real means energy conservation, so we will be doing the right thing anyway in terms of economic policy and environmental policy." A broad array of environmental groups and think tanks, including the Environmental Defense Fund, the Sierra Club, Greenpeace, the World Resources Institute, and the Union of Concerned Scientists, made reductions in carbon-dioxide emissions central to their agendas.

The moral problem seemed clear: human beings were causing the increase of carbon dioxide in the atmosphere. But the moral problem existed only because of a scientific fact—a fact that not only provided justification for doing many of the things that environmentalists wanted to do anyway but also dictated the overriding course of action: reduce carbon-dioxide emissions. Thus science was used to rationalize the moral imperative, unify the environmental agenda, and determine the political solution.

Research as Policy

The summer of 1988 was stultifyingly hot even by Washington, D.C., standards, and the Mississippi River basin was suffering a catastrophic drought. Hansen's proclamation that the greenhouse effect was "changing our climate now" generated a level of public concern sufficient to catch the attention of many politicians. George Bush, who promised to be "the environmental President" and to counter "the greenhouse effect with the White House effect," was elected that November. Despite his campaign rhetoric, the new President was unprepared to offer policies that would curtail fossil-fuel production and consumption or impose economic costs for uncertain political gains. Bush's advisers recognized that support for scientific research offered the best solution politically, because it would give the appearance of action with minimal political risk.

With little debate the Republican Administration and the Democratic Congress in 1990 created the U.S. Global Change Research Program. The program's annual budget reached $1 billion in 1991 and $1.8 billion in 1995, making it one of the largest science initiatives ever undertaken by the U.S. government. Its goal, according to Bush Administration documents, was "to establish the scientific basis for national and international policymaking related to natural and human-induced changes in the global Earth system." A central scientific objective was to "support national and international policymaking by developing the ability to predict the nature and consequences of changes in the Earth system, particularly climate change." A decade and more than $16 billion later, scientific research remains the principal U.S. policy response to climate change.

Meanwhile, the marriage of environmentalism and science gave forth issue: diplomatic efforts to craft a global strategy to reduce carbon-dioxide emissions. Scientists, environmentalists, and government officials, in an attempt to replicate the apparently successful international response to stratospheric-ozone depletion that was mounted in the mid-1980s, created an institutional structure aimed at formalizing the connection between science and political action. The Intergovernmental Panel on Climate Change was established through the United Nations, to provide snapshots of the evolving state of scientific understanding. The IPCC issued major assessments in 1990 and 1996; a third is due early next year. These assessments provide the basis for action under a complementary mechanism, the United Nations Framework Convention on Climate Change. Signed by 154 nations at the 1992 "Earth Summit" in Rio de Janeiro, the convention calls for voluntary reductions in carbon-dioxide emissions. It came into force as an international treaty in March of 1994, and has been ratified by 181 nations. Signatories continue to meet in periodic Conferences of the Parties, of which the most significant to date occurred in Kyoto in 1997, when binding emissions reductions for industrialized countries were proposed under an agreement called the Kyoto Protocol.

The IPCC defines climate change as any sort of change in the earth's climate, no matter what the cause. But the Framework Convention restricts its definition to changes that result from the anthropogenic emission of greenhouse gases. This restriction has profound implications for the framing of the issue. It makes all action under the convention hostage to the ability of scientists not just to document global warming but to attribute it to human causes. An apparently simple question, Are we causing global warming or aren't we?, has become the obsessional focus of science—and of policy.

Finally, if the reduction of carbon-dioxide emissions is an organizing principle for environmentalists, scientists, and environmental-policy makers, it is also an organizing principle for all those whose interests might be threatened by such a reduction. It's easy to be glib about who they might be—greedy oil and coal companies, the rapacious logging industry, recalcitrant automobile manufacturers, corrupt foreign dictatorships—and easy as well to document the excesses and absurdities propagated by some representatives of these groups. Consider, for example, the Greening Earth Society, which "promotes the optimistic scientific view that CO2 is beneficial to humankind and all of nature," and happens to be funded by a coalition of coal-burning utility companies. One of the society's 1999 press releases reported that "there will only be sufficient food for the world's projected population in 2050 if atmospheric concentrations of carbon dioxide are permitted to increase, unchecked." Of course, neither side of the debate has a lock on excess or distortion. The point is simply that the climate-change problem has been framed in a way that catalyzes a determined and powerful opposition.

The Problem With Predictions

When anthropogenic carbon-dioxide emissions became the defining fact for global environmentalism, scientific uncertainty about the causes and consequences of global warming emerged as the apparent central obstacle to action. As we have seen, the Bush Administration justified its huge climate-research initiative explicitly in terms of the need to reduce uncertainty before taking action. Al Gore, by then a senator, agreed, explaining that "more research and better research and better targeted research is absolutely essential if we are going to eliminate the remaining areas of uncertainty and build the broader and stronger political consensus necessary for the unprecedented actions required to address this problem." Thus did a Republican Administration and a Democratic Congress—one side looking for reasons to do nothing, the other seeking justification for action—converge on the need for more research.

How certain do we need to be before we take action? The answer depends, of course, on where our interests lie. Environmentalists can tolerate a good deal more uncertainty on this issue than can, say, the executives of utility or automobile companies. Science is unlikely to overcome such a divergence in interests. After all, science is not a fact or even a set of facts; rather, it is a process of inquiry that generates more questions than answers. The rise in anthropogenic greenhouse-gas emissions, once it was scientifically established, simply pointed to other questions. How rapidly might carbon-dioxide levels rise in the future? How might climate respond to this rise? What might be the effects of that response? Such questions are inestimably complex, their answers infinitely contestable and always uncertain, their implications for human action highly dependent on values and interests.

Having wedded themselves to science, environmentalists must now cleave to it through thick and thin. When research results do not support their cause, or are simply uncertain, they cannot resort to values-based arguments, because their political opponents can portray such arguments as an opportunistic abandonment of rationality. Environmentalists have tried to get out of this bind by invoking the "precautionary principle"—a dandified version of "better safe than sorry"—to advance the idea that action in the presence of uncertainty is justified if potential harm is great. Thus uncertainty itself becomes an argument for action. But nothing is gained by this tactic either, because just as attitudes toward uncertainty are rooted in individual values and interests, so are attitudes toward potential harm.

Charged by the Framework Convention to search for proof of harm, scientists have turned to computer models of the atmosphere and the oceans, called general circulation models, or GCMs.

Carbon-dioxide levels and atmospheric temperatures are measures of the physical state of the atmosphere. GCMs, in contrast, are mathematical representations that scientists use to try to understand past climate conditions and predict future ones. With GCMs scientists seek to explore how climate might respond under different influences—for example, different rates of carbon-dioxide increase. GCMs have calculated global average temperatures for the past century that closely match actual surface-temperature records; this gives climate modelers some confidence that they understand how climate behaves.

Computer models are a bit like Aladdin's lamp—what comes out is very seductive, but few are privy to what goes on inside. Even the most complex models, however, have one crucial quality that non-experts can easily understand: their accuracy can be fully evaluated only after seeing what happens in the real world over time. In other words, predictions of how climate will behave in the future cannot be proved accurate today. There are other fundamental problems with relying on GCMs. The ability of many models to reproduce temperature records may in part reflect the fact that the scientists who designed them already "knew the answer." As John Firor, a former director of the National Center for Atmospheric Research, has observed, climate models "are made by humans who tend to shape or use their models in ways that mirror their own notion of what a desirable outcome would be." Although various models can reproduce past temperature records, and yield similar predictions of future temperatures, they are unable to replicate other observed aspects of climate, such as cloud behavior and atmospheric temperature, and they diverge widely in predicting specific regional climate phenomena, such as precipitation and the frequency of extreme weather events. Moreover, it is simply not possible to know far in advance if the models agree on future temperature because they are similarly right or similarly wrong.

In spite of such pitfalls, a fundamental assumption of both U.S. climate policy and the UN Framework Convention is that increasingly sophisticated models, run on faster computers and supported by more data, will yield predictions that can resolve political disputes and guide action. The promise of better predictions is irresistible to champions of carbon-dioxide reduction, who, after all, must base their advocacy on the claim that anthropogenic greenhouse-gas emissions will be harmful in the future. But regardless of the sophistication of such predictions, new findings will almost inevitably be accompanied by new uncertainties—that's the nature of science—and may therefore act to fuel, rather than to quench, political debate. Our own prediction is that increasingly complex mathematical models that delve ever more deeply into the intricacies and the uncertainties of climate will only hinder political action.

An example of how more scientific research fuels political debate came in 1998, when a group of prominent researchers released the results of a model analyzing carbon-dioxide absorption in North America. Their controversial findings, published in the prestigious journal *Science*, suggested that the amount of carbon dioxide absorbed by U.S. forests might be greater than the amount emitted by the nation's fossil-fuel combustion. This conclusion has two astonishing implications. First, the United States—the world's most profligate energy consumer—may not be directly contributing to rising atmospheric levels of carbon dioxide. Second, the atmosphere seems to be benefiting from young forests in the eastern United States that are particularly efficient at absorbing carbon dioxide. But these young forests exist only because old-growth forests were clear-cut in the eighteenth and nineteenth centuries to make way for farms that were later abandoned in favor of larger, more efficient midwestern farms. In other words, the possibility that the United States is a net carbon-dioxide sink does not reflect efforts to protect the environment; on the contrary, it reflects a history of deforestation and development.

Needless to say, these results quickly made their way into the political arena. At a hearing of the House Resources Committee, Representative John E. Peterson, of Pennsylvania, a Republican, asserted, "There are recent studies that show that in the Northeast, where we have continued to cut timber, and have a regenerating, younger forest, that the greenhouse gases are less when they leave the forest. . . . So a young, growing, vibrant forest is a whole lot better for clean air than an old dying forest."

George Frampton, the director of the White House Council on Environmental Quality, countered, "The science on this needs a lot of work. . . . we need more money for scientific research to under-gird that point of view." How quickly the tables can turn: here was a conservative politician wielding (albeit with limited coherence) the latest scientific results to justify logging old-growth forests in the name of battling global warming, while a Clinton Administration official backpedaled in the manner more typically adopted by opponents of action on climate change—invoking the need for more research.

That's a problem with science—it can turn around and bite you. An even more surprising result has recently emerged from the study of Antarctic glaciers. A strong argument in favor of carbon-dioxide reduction has been the possibility that if temperatures rise owing to greenhouse-gas emissions, glaciers will melt, the sea level will rise, and populous coastal zones all over the world will be inundated. The West Antarctic Ice Sheet has been a subject of particular concern, both because of evidence that it is now retreating and because of geologic studies showing that it underwent catastrophic collapse at least once in the past million years or so. "Behind the reasoned scientific estimates," Greenpeace warns, "lies the possibility of . . . the potential catastrophe of a six meter rise in sea level." But recent research from Antarctica shows that this ice sheet has been melting for thousands of years. Sea-level rise is a problem, but anthropogenic global warming is not the only culprit, and reducing emissions cannot be the only solution.

To make matters more difficult, some phenomena, especially those involving human behavior, are intrinsically unpredictable. Any calculation of future anthropogenic global warming must include an estimate of rates of fossil-fuel combustion in the coming decades. This means that scientists must be able to predict not only the amounts of coal, oil, and natural gas that will be consumed but also changes in the mixture of fossil fuels and other energy sources, such as nuclear, hydro-electric, and solar. These predictions rest on interdependent factors that include energy policies and prices, rates of economic growth, patterns of industrialization and technological innovation, changes in population, and even wars and other geopolitical events. Scientists have no history of being able to predict any of these things. For example, their inability to issue accurate population projections is "one of the best-kept secrets of demography," according to Joel Cohen, the director of the Laboratory of Populations at Rockefeller University. "Most professional demographers no longer believe they can predict precisely the future growth rate, size, composition and spatial distribution of populations," Cohen has observed.

Predicting the human influence on climate also requires an understanding of how climate behaved "normally," before there was any such influence. But what are normal climate patterns? In the absence of human influence, how stationary is climate? To answer such questions, researchers must document and explain the behavior of the pre-industrial climate, and they must also determine how the climate would have behaved over the past two centuries had human beings not been changing the composition of the atmosphere. However, despite the billions spent so far on climate research, Kevin Trenberth, a senior scientist at the National Center for Atmospheric Research, told the *Chicago Tribune* last year, "This may be a shock to many people who assume that we do know adequately what's going on with the climate, but we don't." The National Academy of Sciences reported last year that "deficiencies in the accuracy, quality, and continuity of the [climate] records . . . place serious limitations on the confidence" of research results.

If the normal climate is non-stationary, then the task of identifying the human fingerprint in global climate change becomes immeasurably more difficult. And the idea of a naturally stationary climate may well be chimerical. Climate has changed often and dramatically in the recent past. In the 1940s and 1950s, for example, the East Coast was hammered by a spate of powerful hurricanes, whereas in the 1970s and 1980s hurricanes were much less common. What may appear to be "abnormal" hurricane activity in recent years is abnormal only in relation to this previous quiet period. As far as the ancient climate goes, paleoclimatologists have found evidence of rapid change, even over periods as short as several years. Numerous influences could account for these changes. Ash spewed high into the atmosphere by large volcanoes can reflect solar radiation back into space and result in short-term cooling, as occurred after the 1991 eruption of Mount Pinatubo. Variations in the energy emitted by the sun also affect climate, in ways that are not yet fully understood. Global ocean currents, which move huge volumes of warm and cold water around the world and have a profound influence on climate, can speed up, slow down, and maybe even die out over very short periods of time—perhaps less than a decade. Were the Gulf Stream to shut down, the climate of Great Britain could come to resemble that of Labrador.

Finally, human beings have been changing the surface of the earth for millennia. Scientists increasingly realize that deforestation, agriculture, irrigation, urbanization, and other human activities can lead to major changes in climate on a regional or perhaps even a global scale. Thomas Stohlgren, of the U.S. Geological Survey, has written, "The effects of land use practices on regional climate may overshadow larger-scale temperature changes commonly associated with observed increases in carbon dioxide." The idea that climate may constantly be changing for a variety of reasons does not itself undercut the possibility that anthropogenic carbon dioxide could seriously affect the global climate, but it does confound scientific efforts to predict the consequences of carbon-dioxide emissions.

The Other 80 Percent

If predicting how climate will change is difficult and uncertain, predicting how society will be affected by a changing climate—especially at the local, regional, and national levels, where decision-making takes place—is immeasurably more so. And predicting the impact on climate of reducing carbon-dioxide emissions is so uncertain as to be meaningless. What we do know about climate change suggests that there will be winners and losers, with some areas and nations potentially benefiting from, say, longer growing seasons or more rain, and others suffering from more flooding or drought. But politicians have no way to accurately calibrate the effects-human and economic—of global warming, or the benefits of reducing carbon-dioxide emissions.

Imagine yourself a leading policymaker in a poor, overpopulated, undernourished nation with severe environmental problems. What would it take to get you worried about global warming? You would need to know not just that global warming would make the conditions in your country worse but also that any of the scarce resources you applied to reducing carbon-dioxide emissions would lead to more benefits than if they were applied in another area, such as industrial development or housing construction. Such knowledge is simply unavailable. But you do know that investing in industrial development or better housing would lead to concrete political, economic, and social benefits.

More specifically, suppose that many people in your country live in shacks on a river's floodplain. Floodplains are created and sustained by repeated flooding, so floods are certain to occur in the future, regardless of global warming. Given a choice between building new houses away from the floodplain and converting power plants from cheap local coal to costlier imported fuels, what would you do? New houses would ensure that lives and homes would be saved; a new power plant would reduce carbon-dioxide emissions but leave people vulnerable to floods. In the developing world the carbon-dioxide problem pales alongside immediate environmental and developmental problems. *The China Daily* reported during the 1997 Kyoto Conference:

> The United States . . . and other nations made the irresponsible demand . . . that the developing countries should make commitments to limiting greenhouse gas emissions. . . . As a developing country, China has 60 million poverty-stricken people and China's per capita gas emissions are only one-seventh of the average amount of more developed countries. Ending poverty and developing the economy must still top the agenda of [the] Chinese government.

For the most part, the perspectives of those in the developing world—about 80 percent of the planet's population—have been left outside the frame of the climate-change discussion. This is hardly surprising, considering that the frame was defined mainly by environmentalists and scientists in affluent nations. Developing nations, meanwhile, have quite reasonably refused to agree to the targets for carbon-dioxide reduction set under the Kyoto Protocol. The result may feel like a moral victory to some environmentalists, who reason that industrialized countries, which caused the problem to begin with, should shoulder the primary responsibility for solving it. But the victory is hollow, because most future emissions increases will come from the developing world. In affluent nations almost everyone already owns a full complement of energy-consuming devices. Beyond a certain point increases in income do not result in proportional increases in energy consumption; people simply trade in the old model for a new and perhaps more efficient one. If present trends continue, emissions from the developing world are likely to exceed those from the industrialized nations within the next decade or so.

Twelve years after carbon dioxide became the central obsession of global environmental science and politics, we face the following two realities:

First, atmospheric carbon-dioxide levels will continue to increase. The Kyoto Protocol, which represents the world's best attempt to confront the issue, calls for industrialized nations to reduce their emissions below 1990 levels by the end of this decade. Political and technical realities suggest that not even this modest goal will be achieved. To date, although eighty-four nations have signed the Kyoto Protocol, only twenty-two nations—half of them islands, and none of them major carbon-dioxide emitters—have ratified it. The United States Senate, by a vote of 95-0 in July of 1997, indicated that it would not ratify any climate treaty that lacked provisions requiring developing nations to reduce their emissions. The only nations likely to achieve the emissions commitments set under Kyoto are those, like Russia and Ukraine, whose economies are in ruins. And even successful implementation of the treaty would not halt the progressive increase in global carbon-dioxide emissions.

Second, even if greenhouse-gas emissions could somehow be rolled back to pre-industrial levels, the impacts of climate on society and the environment would continue to increase. Climate affects the world not just through phenomena such as hurricanes and droughts but also because of societal and environmental vulnerability to such phenomena. The horrific toll of Hurricane Mitch reflected not an unprecedented climatic event but a level of exposure typical in developing countries where dense and rapidly increasing populations live in environmentally degraded conditions. Similar conditions underlay more-recent disasters in Venezuela and Mozambique.

If these observations are correct, and we believe they are essentially indisputable, then framing the problem of global warming in terms of carbon-dioxide reduction is a political, environmental, and social dead end. We are not suggesting that humanity can with impunity emit billions of tons of carbon dioxide into the atmosphere each year, or that reducing those emissions is not a good idea. Nor are we making the nihilistic point that since climate undergoes changes for a variety of reasons, there is no need to worry about additional changes imposed by human beings. Rather, we are arguing that environmentalists and scientists, in focusing their own, increasingly congruent interests on carbon-dioxide emissions, have framed the problem of global environmental protection in a way that can offer no realistic prospect of a solution.

Redrawing the Frame

Local weather is the day-to-day manifestation of global climate. Weather is what we experience, and lately there has been plenty to experience. In recent decades human, economic, and environmental losses from disasters related to weather have increased dramatically. Insurance-industry data show that insured losses from weather have been rising steadily. A 1999 study by the German firm Munich Reinsurance Company compared the 1960s with the 1990s and concluded that "the number of great natural catastrophes increased by a factor of three, with economic losses—taking into account the effects of inflation—increasing by a factor of more than eight and insured losses by a factor of no less than sixteen." And yet scientists have been unable to observe a global increase in the number or the severity of extreme weather events. In 1996 the IPCC concluded, "There is no evidence that extreme weather events, or climate variability, has increased, in a global sense, through the 20th century, although data and analyses are poor and not comprehensive."

What has unequivocally increased is society's vulnerability to weather. At the beginning of the twentieth century the earth's population was about 1.6 billion people; today it is about six billion people. Almost four times as many people are exposed to weather today as were a century ago. And this increase has, of course, been accompanied by enormous increases in economic activity, development, infrastructure, and interdependence. In the past fifty years, for example, Florida's population rose fivefold; 80 percent of this burgeoning population lives within twenty miles of the coast. The great Miami hurricane of 1926 made landfall over a small, relatively poor community and caused about $76 million worth of damage (in inflation-adjusted dollars). Today a storm of similar magnitude would strike a sprawling, affluent metropolitan area of two million people, and could cause more than $80 billion worth of damage. The increase in vulnerability is far more dramatic in the developing world, where in an average year tens of thousands of people die in weather-related disasters. According to the *World Disasters Report 1999*, 80 million people were made homeless by weather-related disasters from 1988 to 1997. As the population and vulnerability of the developing world continue to rise, such numbers will continue to rise as well, with or without global warming.

Environmental vulnerability is also on the rise. The connections between weather impacts and environmental quality are immediate and obvious—much more so than the connections between global warming and environmental quality. Deforestation, the destruction of wetlands, and the development of fragile coastlines can greatly magnify flooding; floods, in turn, can mobilize toxic chemicals in soil and storage facilities and cause devastating pollution of water sources and harm to wildlife. Poor agricultural, forest-management, and grazing practices can exacerbate the effects of drought, amplify soil erosion, and promote the spread of wildfires. Damage to the environment due to deforestation directly contributed to the devastation wrought by Hurricane Mitch, as denuded hillsides washed away in catastrophic landslides, and excessive development along unmanaged floodplains put large numbers of people in harm's way.

Our view of climate and the environment draws on people's direct experience and speaks to widely shared values. It therefore has an emotional and moral impact that can translate into action. This view is framed by four precepts. First, the impacts of weather and climate are a serious threat to human welfare in the present and are likely to get worse in the future. Second, the only way to reduce these impacts is to reduce societal vulnerability to them. Third, reducing vulnerability can be achieved most effectively by encouraging democracy, raising standards of living, and improving environmental quality in the developing world. Fourth, such changes offer the best prospects not only for adapting to a capricious climate but also for reducing carbon-dioxide emissions.

The implicit moral imperative is not to prevent human disruption of the environment but to ameliorate the social and political conditions that lead people to behave in environmentally disruptive ways. This is a critical distinction—and one that environmentalists and scientists embroiled in the global-warming debate have so far failed to make.

To begin with, any global effort to reduce vulnerability to weather and climate must address the environmental conditions in developing nations. Poor land-use and natural-resource-management practices are, of course, a reflection of poverty, but they are also caused by government policies, particularly those that encourage unsustainable environmental activities. William Ascher, a political scientist at Duke University, has observed that such policies typically do not arise out of ignorance or lack of options but reflect conscious tradeoffs made by government officials faced with many competing priorities and political pressures. Nations, even poor ones, have choices. It was not inevitable, for example, that Indonesia would promote the disastrous exploitation of its forests by granting subsidized logging concessions to military and business leaders. This was the policy of an autocratic government seeking to manipulate powerful sectors of society. In the absence of open, democratically responsive institutions, Indonesian leaders were not accountable for the costs that the public might bear, such as increased vulnerability to floods, landslides, soil erosion, drought, and fire. Promoting democratic institutions in developing nations could be the most important item on an agenda aimed at protecting the global environment and reducing vulnerability to climate. Environmental groups concerned about the consequences of climate change ought to consider reorienting their priorities accordingly.

Such long-term efforts must be accompanied by activities with a shorter-term payoff. An obvious first step would be to correct some of the imbalances created by the obsession with carbon dioxide. For example, the U.S. Agency for International Development has allocated $1 billion over five years to help developing nations quantify, monitor, and reduce greenhouse-gas emissions, but is spending less than a tenth of that amount on programs to prepare for and prevent disasters. These priorities should be rearranged. Similarly, the United Nations' International Strategy for Disaster Reduction is a relatively low-level effort that should be elevated to a status comparable to that of the Framework Convention on Climate Change.

Intellectual and financial resources are also poorly allocated in the realm of science, with research focused disproportionately on understanding and predicting basic climatic processes. Such research has yielded much interesting information about the global climate system. But little priority is given to generating and disseminating knowledge that people and communities can use to reduce their vulnerability to climate and extreme weather events. For example, researchers have made impressive strides in anticipating the impacts of some relatively short-term climatic phenomena, notably El Niño and La Niña. If these advances were accompanied by progress in monitoring weather, identifying vulnerable regions and populations, and communicating useful information, we would begin to reduce the toll exacted by weather and climate all over the world.

A powerful international mechanism for moving forward already exists in the Framework Convention on Climate Change. The language of the treaty offers sufficient flexibility for new priorities. The text states that signatory nations have an obligation to "cooperate in preparing for adaptation to the impacts of climate change [and to] develop and elaborate appropriate and integrated plans for coastal zone management, water resources and agriculture, and for the protection and rehabilitation of areas . . . affected by drought and desertification, as well as floods."

The idea of improving our adaptation to weather and climate has been taboo in many circles, including the realms of international negotiation and political debate. "Do we have so much faith in our own adaptability that we will risk destroying the integrity of the entire global ecological system?" Vice President Gore asked in his book *Earth in the Balance* (1992). "Believing that we can adapt to just about anything is ultimately a kind of laziness, an arrogant faith in our ability to react in time to save our skin." For environmentalists, adaptation represents a capitulation to the momentum of human interference in nature. For their opponents, putting adaptation on the table would mean acknowledging the reality of global warming. And for scientists, focusing on adaptation would call into question the billions of tax dollars devoted to research and technology centered on climate processes, models, and predictions.

Yet there is a huge potential constituency for efforts focused on adaptation: everyone who is in any way subject to the effects of weather. Reframing the climate problem could mobilize this constituency and revitalize the Framework Convention. The revitalization could concentrate on coordinating disaster relief, debt relief, and development assistance, and on generating and providing information on climate that participating countries could use in order to reduce their vulnerability.

An opportunity to advance the cause of adaptation is on the horizon. The U.S. Global Change Research Program is now finishing its report on the National Assessment of the Potential Consequences of Climate Variability and Change. The draft includes examples from around the United States of why a greater focus on adaptation to climate makes sense. But it remains to be seen if the report will redefine the terms of the climate debate, or if it will simply become fodder in the battle over carbon-dioxide emissions.

Finally, efforts to reduce carbon-dioxide emissions need not be abandoned. The Framework Convention and its offshoots also offer a promising mechanism for promoting the diffusion of energy-efficient technologies that would reduce emissions. Both the convention and the Kyoto Protocol call on industrialized nations to share new energy technologies with the developing world. But because these provisions are coupled to carbon-dioxide-reduction mandates, they are trapped in the political grid-lock. They should be liberated, promoted independently on the basis of their intrinsic environmental and economic benefits, and advanced through innovative funding mechanisms. For example, as the United Nations Development Programme has suggested, research into renewable-energy technologies for poor countries could be supported in part by a modest levy on patents registered under the World Intellectual Property Organization. Such ideas should be far less divisive than energy policies advanced on the back of the global-warming agenda.

As an organizing principle for political action, vulnerability to weather and climate offers everything that global warming does not: a clear, uncontroversial story rooted in concrete human experience, observable in the present, and definable in terms of unambiguous and widely shared human values, such as the fundamental rights to a secure shelter, a safe community, and a sustainable environment. In this light, efforts to blame global warming for extreme weather events seem maddeningly perverse—as if to say that those who died in Hurricane Mitch were symbols of the profligacy of industrialized society, rather than victims of poverty and the vulnerability it creates.

Such perversity shows just how morally and politically dangerous it can be to elevate science above human values. In the global-warming debate the logic behind public discourse and political action has been precisely backwards. Environmental prospects for the coming century depend far less on our strategies for reducing carbon-dioxide emissions than on our determination and ability to reduce human vulnerability to weather and climate.

Chapter 2 ■ Breaking the Global Warming Gridlock

Questions for Discussion

1. How much is decision making guided by paradigms and how much by politics? Or, should we ask the question this way: how often are paradigms based less on science than on ideology? What do Sarewitz and Pielke suggest? What implications does this have for your own project?

2. What do you make of the argument presented by Sarewitz and Pielke for breaking the "global warming gridlock"? Do you think their argument presents a successful strategy for creating change? Would you call it a potential "paradigm shift"? What do current events suggest about how well their approach might work, or how much it might shape a new consensus?

Absolute PowerPoint

Can a software package edit our thoughts?

Ian Parker

■ ■ ■

Before there were presentations, there were conversations, which were a little like presentations but used fewer bullet points, and no one had to dim the lights. A woman we can call Sarah Wyndham, a defense-industry consultant living in Alexandria, Virginia, recently began to feel that her two daughters weren't listening when she asked them to clean their bedrooms and do their chores. So, one morning, she sat down at her computer, opened Microsoft's PowerPoint program, and typed:

FAMILY MATTERS
An approach for positive change
to the Wyndham family team.

On a new page, she wrote:

- Lack of organization leads to confusion and frustration among family members.

- Disorganization is detrimental to grades and to your social life.

- Disorganization leads to inefficiencies that impact the entire family.

Instead of pleading for domestic harmony, Sarah Wyndham was pitching for it. Soon she had eighteen pages of large type, supplemented by a color photograph of a generic happy family riding bicycles, and, on the final page, a drawing of a key—the key to success. The briefing was given only once, last fall. The experience was so upsetting to her children that the threat of a second showing was enough to make one of the Wyndham girls burst into tears.

PowerPoint, which can be found on two hundred and fifty million computers around the world, is software you impose on other people. It allows you to arrange text and graphics in a series of pages, which you can project, slide by slide, from a laptop computer onto a screen, or print as a booklet (as Sarah Wyndham did). The usual metaphor for everyday software is the tool, but that doesn't seem to be right here. PowerPoint is more like a suit of clothes, or a car, or plastic surgery. You take it out with you. You are judged by it—you insist on being judged by it. It is by definition a social instrument, turning middle managers into bullet-point dandies.

But PowerPoint also has a private, interior influence. It edits ideas. It is, almost surreptitiously, a business manual as well as a business suit, with an opinion—an oddly pedantic, prescriptive opinion—about the way we should think. It helps you make a case, but it also makes its own case: about how to organize information, how much information to organize, how to look at the world. One feature of this is the AutoContent Wizard, which supplies templates—"Managing Organizational Change" or

"Communicating Bad News," say—that are so close to finished presentations you barely need to do more than add your company logo. The "Motivating a Team" template, for example, includes a slide headed "Conduct a Creative Thinking Session:"

Ask: In what ways can we. . . . ?

- Assess the situation. Get the facts.
- Generate possible solutions with green light, nonjudgmental thinking.
- Select the best solution.

The final injunction is "Have an inspirational close."

It's easy to avoid these extreme templates—many people do—as well as embellishments like clip art, animations, and sound effects. But it's hard to shake off AutoContent's spirit: even the most easy going PowerPoint template insists on a heading followed by bullet points, so that the user is shepherded toward a staccato, summarizing frame of mind, of the kind parodied, for example, in a PowerPoint Gettysburg Address posted on the Internet: "Dedicate portion of field—fitting!"

Because PowerPoint can be an impressive antidote to fear—converting public-speaking dread into moviemaking pleasure—there seems to be no great impulse to fight this influence, as you might fight the unrelenting animated paperclip in Microsoft Word. Rather, PowerPoint's restraints seem to be soothing—so much so that where Microsoft has not written rules, businesses write them for themselves. A leading U.S. computer manufacturer has distributed guidelines to its employees about PowerPoint presentations, insisting on something it calls the "Rule of Seven": "Seven (7) bullets or lines per page, seven (7) words per line."

Today, after Microsoft's decade of dizzying growth, there are great tracts of corporate America where to appear at a meeting without PowerPoint would be unwelcome and vaguely pretentious, like wearing no shoes. In darkened rooms at industrial plants and ad agencies, at sales pitches and conferences, this is how people are communicating: no paragraphs, no pronouns—the world condensed into a few upbeat slides, with seven or so words on a line, seven or so lines on a slide. And now it's happening during sermons and university lectures and family arguments, too. A New Jersey PowerPoint user recently wrote in an online discussion, "Last week I caught myself planning out (in my head) the slides I would need to explain to my wife why we couldn't afford a vacation this year." Somehow, a piece of software designed, fifteen years ago, to meet a simple business need has become a way of organizing thought at kindergarten show-and-tells. "Oh, Lord," one of the early developers said to me. "What have we done?"

Forty years ago, a workplace meeting was a discussion with your immediate colleagues. Engineers would meet with other engineers and talk in the language of engineering. A manager might make an appearance—acting as an interpreter, a bridge to the rest of the company—but no one from the marketing or production or sales department would be there. Somebody might have gone to the trouble of cranking out mimeographs—that would be the person with purple fingers.

But the structure of American industry changed in the nineteen-sixties and seventies. Clifford Nass, who teaches in the Department of Communication at Stanford, says, "Companies weren't discovering things in the laboratory and then trying to convince consumers to buy them. They were discovering—or creating—consumer demand, figuring out what they can convince consumers they need, then going to the laboratory and saying, 'Build this!' People were saying, 'We can create demand. Even if demand doesn't exist, we know how to market this.' SpaghettiOs is the great example. The guy came up with the jingle first: 'The neat round spaghetti you can eat with a spoon.' And he said, 'Hey! Make spaghetti in the shape of small circles!' "

As Jerry Porras, a professor of organizational behavior and change at Stanford Graduate School of Business, says, "When technologists no longer just drove the product out but the customer sucked it out, then you had to know what the customer wanted, and that meant a lot more interaction inside the company." There are new conversations: Can we make this? How do we sell this if we make it? Can we do it in blue?

America began to go to more meetings. By the early nineteen-eighties, when the story of PowerPoint starts, employees had to find ways to talk to colleagues from other departments, colleagues who spoke a different language, brought together by SpaghettiOs and by the simple fact that technology was generating more information. There was more to know and, as the notion of a job for life eroded, more reason to know it.

In this environment, visual aids were bound to thrive. In 1975, fifty thousand overhead projectors were sold in America. By 1985, that figure had increased to more than a hundred and twenty thousand. Overheads, which were developed in the mid-forties for use by the police, and were then widely used in bowling alleys and schools, did not fully enter business life until the mid-seventies, when a transparency film that could survive the heat of a photocopier became available. Now anything on a sheet of paper could be transferred to an overhead slide. Overheads were cheaper than the popular alternative, the 35-mm. slide (which needed graphics professionals), and they were easier to use. But they restricted you to your typewriter's font—rather, your secretary's typewriter's font—or your skill with Letraset and a felt-tipped pen. A businessman couldn't generate a handsome, professional-looking font in his own office.

In 1980, though, it was clear that a future of widespread personal computers—and laser printers and screens that showed the very thing you were about to print—was tantalizingly close. In the Mountain View, California, laboratory of Bell-Northern Research, computer-research scientists had set up a great mainframe computer, a graphics workstation, a phototypesetter, and the earliest Canon laser printer, which was the size of a bathtub and took six men to carry into the building—together, a cumbersome approximation of what would later fit on a coffee table and cost a thousand dollars. With much trial and error, and jogging from one room to another, you could use this collection of machines as a kind of word processor.

Whitfield Diffie had access to this equipment. A mathematician, a former peacenik, and an enemy of exclusive government control of encryption systems, Diffie had secured a place for himself in computing legend in 1976, when he and a colleague, Martin Hellman, announced the discovery of a new method of protecting secrets electronically—public-key cryptography. At Bell-Northern, Diffie was researching the security of telephone systems. In 1981, preparing to give a presentation with 35-mm. slides, he wrote a little program, tinkering with some graphics software designed by a B.N.R. colleague, that allowed you to draw a black frame on a piece of paper. Diffie expanded it so that the page could show a number of frames, and text inside each frame, with space for commentary around them. In other words, he produced a storyboard—a slide show on paper—that could be sent to the designers who made up the slides, and that would also serve as a script for his lecture. (At this stage, he wasn't photocopying what he had produced to make overhead transparencies, although scientists in other facilities were doing that.) With a few days' effort, Diffie had pointed the way to PowerPoint.

Diffie has long gray hair and likes to wear fine English suits. Today, he works for Sun Microsystems, as an internal consultant on encryption matters. I recently had lunch with him in Palo Alto, and for the first time he publicly acknowledged his presence at the birth of PowerPoint. It was an odd piece of news: as if Lenin had invented the stapler. Yes, he said, PowerPoint was "based on" his work at B.N.R. This is not of great consequence to Diffie, whose reputation in his own field is so high that he is one of the few computer scientists to receive erotically charged fan mail. He said he was "mildly miffed" to have made no money from the PowerPoint connection, but he has no interest in beginning a feud with an old friend. "Bob was the one who had the vision to understand how important it was to the world," he said. "And I didn't."

Bob is Bob Gaskins, the man who has to take final responsibility for the drawn blinds of high-rise offices around the world and the bullet points dashing across computer screens inside. His account of PowerPoint's parentage does not exactly match Diffie's, but he readily accepts his former colleague as "my inspiration." In the late nineteen-seventies and early eighties, Gaskins was B.N.R.'s head of computer-science research. A former Berkeley Ph.D. student, he had a family background in industrial photographic supplies and grew up around overhead projectors and inks and gels. In 1982, he returned from a six-month overseas business trip and, with a vivid sense of the future impact of the Apple Macintosh and of Microsoft's Windows (both of which were in development), he wrote a list of fifty commercial possibilities—Arabic typesetting, menus, signs. And then he looked around his own laboratory and realized what had happened while he was away: following Diffie's lead, his colleagues were trying to make overheads to pitch their projects for funding, despite the difficulties of using the equipment. (What you saw was not at all what you got.) "Our mainframe was buckling under the load," Gaskins says.

He now had his idea: a graphics program that would work with Windows and the Macintosh, and that would put together, and edit, a string of single pages, or "slides." In 1984, he left B.N.R., joined an ailing Silicon Valley software firm, Forethought, in exchange for a sizable share of the company, and hired a software developer, Dennis Austin. They began work on a program called Presenter. After a trademark problem, and an epiphany Gaskins had in the shower, Presenter became PowerPoint.

Gaskins is a precise, bookish man who lives with his wife in a meticulously restored and furnished nineteenth-century house in the Fillmore district of San Francisco. He has recently discovered an interest in antique concertinas. When I visited him, he was persuaded to play a tune, and he gave me a copy of a forthcoming paper he had co-written: "A Wheatstone Twelve-Sided 'Edeophone' Concertina with Pre-MacCann Chromatic Duet Fingering." Gaskins is skeptical about the product that PowerPoint has become—AutoContent and animated fades between slides—but he is devoted to the simpler thing that it was, and he led me through a well-preserved archive of PowerPoint memorabilia, including the souvenir program for the PowerPoint reunion party, in 1997, which had a quiz filled with in-jokes about font size and programming languages. He also found an old business plan from 1984. One phrase—the only one in italics—read, "Allows the content-originator to control the presentation." For Gaskins, that had always been the point: to get rid of the intermediaries—graphic designers—and never mind the consequences. Whenever colleagues sought to restrict the design possibilities of the program (to make a design disaster less likely), Gaskins would overrule them, quoting Thoreau: "I came into this world, not chiefly to make this a good place to live in, but to live in it, be it good or bad."

PowerPoint 1.0 went on sale in April, 1987—available only for the Macintosh, and only in black-and-white. It generated text-and-graphics pages that a photocopier could turn into overhead transparencies. (This was before laptop computers and portable projectors made PowerPoint a tool for live electronic presentations. Gaskins thinks he may have been the first person to use the program in the modern way, in a Paris hotel in 1992—which is like being the first person ever to tap a microphone and say, "Can you hear me at the back?") The Macintosh market was small and specialized, but within this market PowerPoint—the first product of its kind—was a hit. "I can't describe how wonderful it was," Gaskins says. "When we demonstrated at tradeshows, we were mobbed." Shortly after the launch, Forethought accepted an acquisition offer of fourteen million dollars from Microsoft. Microsoft paid cash and allowed Bob Gaskins and his colleagues to remain partly self-governing in Silicon Valley, far from the Microsoft campus, in Redmond, Washington. Microsoft soon regretted the terms of the deal; PowerPoint workers became known for a troublesome independence of spirit (and for rewarding themselves, now and then, with beautifully staged parties—caviar, string quartets, Renaissance-period fancy dress).

PowerPoint had been created, in part, as a response to the new corporate world of interdepartmental communication. Those involved with the program now experienced the phenomenon at first hand. In 1990, the first PowerPoint for Windows was launched, alongside Windows 3.0. And PowerPoint quickly became what Gaskins calls "a cog in the great machine." The PowerPoint programmers were forced to make unwelcome changes, partly because in 1990 Word, Excel, and PowerPoint began to be integrated into Microsoft Office—a strategy that would eventually make PowerPoint invincible—and partly in response to market research. AutoContent was added in the mid-nineties, when Microsoft learned that some would-be presenters were uncomfortable with a blank PowerPoint page—it was hard to get started. "We said, 'What we need is some automatic content!'" a former Microsoft developer recalls, laughing. " 'Punch the button and you'll have a presentation.' " The idea, he thought, was "crazy." And the name was meant as a joke. But Microsoft took the idea and kept the name—a rare example of a product named in outright mockery of its target customers.

Gaskins left PowerPoint in 1992, and many of his colleagues followed soon after. Now rich from Microsoft stock, and beginning the concertina-collecting phase of their careers, they watched as their old product made its way into the heart of American business culture. By 1993, PowerPoint had a majority share of the presentation market. In 1995, the average user created four and a half presentations a month. Three years later, the monthly average was nine. PowerPoint began to appear in cartoon strips and everyday conversation. A few years ago, Bob Gaskins was at a presentations-heavy conference in Britain. The organizer brought the proceedings to a sudden stop, saying, "I've just been told that the inventor of PowerPoint is in the audience—will he please identify himself so we can recognize his contribution to the advancement of science?" Gaskins stood up. The audience laughed and applauded.

Cathleen Belleville, a former graphic designer who worked at PowerPoint as a product planner from 1989 to 1995, was amazed to see a clip-art series she had created become modern business icons. The images were androgynous silhouette stick figures (she called them Screen Beans), modeled on a former college roommate: a little figure clicking its heels; another with an inspirational light bulb above its head. One Screen Bean, the patron saint of PowerPoint—a figure that stands beneath a question mark, scratching its head in puzzlement—is so popular that a lawyer at a New York firm who has seen many PowerPoint presentations claims never to have seen one without the head-scratcher. Belleville herself has seen her Beans all over the world, reprinted on baseball caps, blown up fifteen feet high in a Hamburg bank. "I told my mom, 'You know, my artwork is in danger of being more famous than the "Mona Lisa." '" Above the counter in a laundromat on Third Avenue in New York, a sign explains that no responsibility can be taken for deliveries to doorman buildings. And there, next to the words, is the famous puzzled figure. It is hard to understand the puzzlement. Doorman? Delivery? But perhaps this is simply how a modern poster clears its throat: Belleville has created the international sign for "sign."

According to Microsoft estimates, at least thirty million PowerPoint presentations are made every day. The program has about ninety-five per cent of the presentations-software market. And so perhaps it was inevitable that it would migrate out of business and into other areas of our lives. I recently spoke to Sew Meng Chung, a Malaysian research engineer living in Singapore who got married in 1999. He told me that, as his guests took their seats for a wedding party in the Goodwood Park Hotel, they were treated to a PowerPoint presentation: a hundred and thirty photographs—one fading into the next every four or five seconds, to musical accompaniment. "They were baby photos, and courtship photos, and photos taken with our friends and family," he told me.

I also spoke to Terry Taylor, who runs a website called eBibleTeacher.com, which supplies materials for churches that use electronic visual aids. "Jesus was a storyteller, and he gave graphic images," Taylor said. "He would say, 'Consider the lilies of the field, how they grow,' and all indications are that there were lilies in the field when he was talking, you know. He used illustrations." Taylor estimates that fifteen per cent of American churches now have video projectors, and many use PowerPoint regularly for announcements, for song lyrics, and to accompany preaching. (Taylor has seen more than one sermon featuring the head-scratching figure.) Visitors to Taylor's site can download photographs of locations in the Holy Land, as well as complete PowerPoint sermons—for example, "Making Your Marriage Great:"

Find out what you are doing to harm your marriage and heal it!

- Financial irresponsibility.

- Temper.

- Pornography.

- Substance abuse.

- You name it!

When PowerPoint is used to flash hymn lyrics, or make a quick pitch to a new client, or produce an eye-catching laundromat poster, it's easy to understand the enthusiasm of, say, Tony Kurz, the vice-president for sales and marketing of a New York-based Internet company, who told me, "I love PowerPoint. It's a brilliant application. I can take you through at exactly the pace I want to take you." There are probably worse ways to transmit fifty or a hundred words of text, or information that is mainly visual—ways that involve more droning, more drifting. And PowerPoint demands at least some rudimentary preparation: a PowerPoint presenter is, by definition, not thinking about his or her material for the very first time. Steven Pinker, the author of "The Language Instinct" and a psychology professor at the Massachusetts Institute of Technology, says that PowerPoint can give visual shape to an argument. "Language is a linear medium: one damn word after another," he says. "But ideas are multidimensional. . . . When properly employed, PowerPoint makes the logical structure of an argument more transparent. Two channels sending the same information are better than one."

Still, it's hard to be perfectly comfortable with a product whose developers occasionally find themselves trying to suppress its use. Jolene Rocchio, who is a product planner for Microsoft Office (and is upbeat about PowerPoint in general), told me that, at a recent meeting of a nonprofit organization in San Francisco, she argued against a speaker's using PowerPoint at a future conference. "I said, 'I think we just need her to get up and speak.' " On an earlier occasion, Rocchio said, the same speaker had tried to use PowerPoint and the projector didn't work, "and everybody was, like, cheering. They just wanted to hear this woman speak, and they wanted it to be from her heart. And the PowerPoint almost alienated her audience."

This is the most common complaint about PowerPoint. Instead of human contact, we are given human display. "I think that we as a people have become unaccustomed to having real conversations with each other, where we actually give and take to arrive at a new answer. We present to each other, instead of discussing," Cathy Belleville says. Tad Simons, the editor of the magazine *Presentations* (whose second-grade son used PowerPoint for show-and-tell), is familiar with the sin of triple delivery, where precisely the same text is seen on the screen, spoken aloud, and printed on the handout in front of you (the "leave-behind," as it is known in some circles). "The thing that makes my heart sing is when somebody presses the 'B' button and the screen goes black and you can actually talk to the person," Simons told me.

In 1997, Sun Microsystems' chairman and C.E.O., Scott McNealy, "banned" PowerPoint (a ban widely disregarded by his staff). The move might have been driven, in part, by Sun's public-relations needs as a Microsoft rival, but, according to McNealy, there were genuine productivity issues. "Why did we ban it? Let me put it this way: If I want to tell my forty thousand employees to attack, the word 'attack' in ASCII is forty-eight bits. As a Microsoft Word document, it's 90,112 bits. Put that same word in a PowerPoint slide and it becomes 458,048 bits. That's a pig through the python when you try to send it over the Net." McNealy's concern is shared by the American military. Enormously elaborate PowerPoint files (generated by presentation-obsessives—so-called PowerPoint Rangers) were said to be clogging up the military's bandwidth. Last year, to the delight of many under his command,

General Henry H. Shelton, the chairman of the Joint Chiefs of Staff, issued an order to U.S. bases around the world insisting on simpler presentations.

PowerPoint was developed to give public speakers control over design decisions. But it's possible that those speakers should be making other, more important decisions. "In the past, I think we had an inefficient system, where executives passed all of their work to secretaries," Cathy Belleville says. "But now we've got highly paid people sitting there formatting slides—spending hours formatting slides—because it's more fun to do that than concentrate on what you're going to say. It would be much more efficient to offload that work onto someone who could do it in a tenth of the time, and be paid less. Millions of executives around the world are sitting there going, `Arial? Times Roman? Twenty-four point? Eighteen point?' "

In the glow of a PowerPoint show, the world is condensed, simplified, and smoothed over—yet bright and hyperreal—like the cityscape background in a PlayStation motor race. PowerPoint is strangely adept at disguising the fragile foundations of a proposal, the emptiness of a business plan; usually, the audience is respectfully still (only venture capitalists dare to dictate the pace of someone else's slide show), and, with the visual distraction of a dancing pie chart, a speaker can quickly move past the laughable flaw in his argument. If anyone notices, it's too late—the narrative presses on.

Last year, three researchers at Arizona State University, including Robert Cialdini, a professor and the author of "Influence: Science and Practice," conducted an experiment in which they presented three groups of volunteers with information about Andrew, a fictional high-school student under consideration for a university football scholarship. One group was given Andrew's football statistics typed on a piece of paper. The second group was shown bar graphs. Those in the third group were given a PowerPoint presentation, in which animated bar graphs grew before their eyes.

Given Andrew's record, what kind of prospect was he? According to Cialdini, when Andrew was PowerPointed, viewers saw him as a greater potential asset to the football team. The first group rated Andrew four and a half on a scale of one to seven; the second rated him five; and the PowerPoint group rated him six. PowerPoint gave him power. The experiment was repeated, with three groups of sports fans that were accustomed to digesting sports statistics; this time, the first two groups gave Andrew the same rating. But the group that saw the PowerPoint presentation still couldn't resist it. Again, Andrew got a six. PowerPoint seems to be a way for organizations to turn expensive, expert decision-makers into novice decision-makers. "It's frightening," Cialdini says. He always preferred to use slides when he spoke to business groups, but one high-tech company recently hinted that his authority suffered as a result. "They said, 'You know what, Bob? You've got to get into PowerPoint, otherwise people aren't going to respond.' So I made the transfer."

Clifford Nass has an office overlooking the Oval lawn at Stanford, a university where the use of PowerPoint is so widespread that to refrain from using it is sometimes seen as a mark of seniority and privilege, like egg on one's tie. Nass once worked for Intel, and then got a Ph.D. in sociology, and now he writes about and lectures on the ways people think about computers. But, before embarking on any of that, Professor Nass was a professional magician—Cliff Conjure—so he has some confidence in his abilities as a public performer.

According to Nass, who now gives PowerPoint lectures because his students asked him to, PowerPoint "lifts the floor" of public speaking: a lecture is less likely to be poor if the speaker is using the program. "What PowerPoint does is very efficiently deliver content," Nass told me. "What students gain is a lot more information—not just facts but rules, ways of thinking, examples." At the same time, PowerPoint "lowers the ceiling," Nass says. "What you miss is the process. The classes I remember most, the professors I remember most, were the ones where you could watch how they thought. You don't remember what they said, the details. It was 'What an elegant way to wrap around a problem!' PowerPoint takes that away. PowerPoint gives you the outcome, but it removes the process."

"What I miss is, when I used to lecture without PowerPoint, every now and then I'd get a cool idea," he went on. "I remember once it just hit me. I'm lecturing, and all of a sudden I go, 'God! "The Wizard of Oz"! The scene at the end of "The Wizard of Oz"!" Nass, telling this story, was almost shouting. (The lecture, he later explained, was about definitions of "the human" applied to computers.) "I just went for it—twenty-five minutes. And to this day students who were in that class remember it. That couldn't happen now: 'Where the hell is the slide?'

PowerPoint could lead us to believe that information is all there is. According to Nass, PowerPoint empowers the provider of simple content (and that was the task Bob Gaskins originally set for it), but it risks squeezing out the provider of process—that is to say, the rhetorician, the storyteller, the poet, the person whose thoughts cannot be arranged in the shape of an AutoContent slide. "I hate to admit this," Nass said, "but I actually removed a book from my syllabus last year because I couldn't figure out how to PowerPoint it. It's a lovely book called 'Interface Culture,' by Steven Johnson, but it's very discursive; the charm of it is the throwaways. When I read this book, I thought, "My head's filled with ideas, and now I've got to write out exactly what those ideas are, and—they're not neat." He couldn't get the book into bullet points; every time he put something down, he realized that it wasn't quite right. Eventually, he abandoned the attempt, and, instead of a lecture, he gave his students a recommendation. He told them it was a good book, urged them to read it, and moved on to the next bullet point.

Chapter 2 ■ Absolute PowerPoint

Questions for Discussion

1. Throughout his article, Parker is critical of the way PowerPoint has provided so many pre-formatted presentations through the AutoContent Wizard. Why have business people found this a useful feature of the program? And why does this trouble Parker? What larger problem does he see growing out of PowerPoint's cultural hegemony? And are people in the business world more or less likely to agree with him over time?

2. Do you find Parker's critique of the AutoContent Wizard persuasive? Is it likely to affect the way you use PowerPoint yourself? If yes, how so? For example, are you now more or less likely to use the Screen Beans cartoon characters?

3. Parker's history of PowerPoint can be read as a history of "paradigm formation" (or "paradigm shift") within organizational culture. What drove that shift? What part did technology play in the change? What factors had to come together to make PowerPoint possible? How can it serve as a model for developing other successful products?

4. Parker is especially critical of the culture of active presentation and passive reception that PowerPoint breeds, in his view. Why is this a problem, for example, in a college course? Why should it trouble us, according to Parker, that people seem to deliver and attend more "presentations" than "conversations"? What is the difference between the two, and why is it important to Parker? Do you agree with the assumptions about the way the world should be that his distinction suggests?

5. The experiment involving the high-school student "Andrew" as the subject of a presentation seems an especially interesting example in Parker's article. What conclusions are we expected to draw from this example? How might this example affect the way you watch other students' presentations in this course?

The Résumé
and Cover Letter

Chapter

The Assignment

Prepare a résumé and cover letter in response to a specific, published job posting or advertisement. I recommend you use a posting in a newspaper or on the Internet so that you can offer a print copy. You must bring in a copy of the job listing for peer revision day and hand it in with your final assignment.

Unless your instructor chooses to set forth more specific guidelines, here is the assignment:

- Each document must be one page only (unless your instructor indicates otherwise).

- You must turn in the job announcement with your assignment or it is incomplete.

- These documents should be prepared according to standards discussed in class.

- They should be proofread closely so that there are no errors in either document.

- The assignment and any drafts discussed in class must be turned in according to the schedule set by your instructor. If you fail to turn in the assignment on time or if you fail to have a sufficient draft for peer revision, it will affect your grade for the assignment.

Students must bring the job advertisement on the day of peer revision, since without it peers and instructors cannot judge audience expectations. The résumé should be ordered in a way that best responds to the potential employer's needs, and the cover letter should offer significant details distinguishing the candidate and highlighting aspects of the résumé in a way that clearly responds to those needs. The cover letter should offer a high level of detail and should interpret the résumé for the potential employer.

I am always surprised by the level of error on the résumé, which ought to have absolutely no errors of syntax, grammar, consistency, or sense. Errors in consistency (in spacing, parallel form, layout, and capitalization) are especially prevalent. General sloppiness or failure to adhere to accepted principles (such as using active verbs) will definitely factor into your grade.

Sample Résumés and Cover Letters

The following comments and questions will help you think about the sample résumés and cover letters.

Joel Anderson

Job Advertisement:

INTERNSHIP OPPORTUNITY: Northwestern Mutual Financial Network is looking for interns to work in our Trust Services division. A working knowledge of GAAP and IFRS required. Personal finance experience preferred. Individuals who work well with peers and have leadership potential are encouraged to apply.

After meeting Sara Creighton at a career fair, Joel Anderson submitted this cover letter and résumé. Notice how Joel followed recommended guidelines in organizing his cover letter. The introductory paragraph identifies the position sought and forecasts the letter's content. The second paragraph discusses Joel's qualifications for the internship position. The final paragraph requests an interview and states Joel's phone number and e-mail address. Likewise, Joel's résumé is clearly organized and balanced in format.

However, upon closer examination, we see the substance of Joel's letter and résumé require revision. How might Joel better tailor his letter to his audience? How could Joel more accurately forecast the key points of his letter in the introductory paragraph? How could he be more specific when describing the skills he has developed? Where would transitions help to better show the logic between his ideas? What unnecessary words could Joel eliminate? Which skills do you confidently believe Joel has? Why? How might the concluding paragraph more politely and confidently request an interview? What revisions would you suggest to improve Joel's résumé to make it clearer and more specific? Remember that details lead the reader to conclude that the writer is a terrific job candidate; telling the reader "I am terrific" is unconvincing.

Joan Parker

Job Advertisement:

INDUSTRIAL ENGINEER INTERNSHIP: Seeking high-achieving engineering majors to work as summer interns for credit only. Gain valuable hands-on experience at a busy engineering firm. Strong clerical skills and knowledge of design software required.

Joan's experience seems appropriate to the internship. But has Joan's letter emphasized her most relevant experience? What could the writer have done to be more specific in the letter? What abilities is the employer most likely looking for in a job candidate? How might Joan better tailor her letter to highlight these abilities? How could the résumé be improved?

Lynn Cato

Job Advertisement:

VETERINARY ASSISTANT: Part-time, flexible hours. Must be reliable, self-motivated, people and animal friendly. Good experience for pre-vet students. Call us at 732-555-1234.

Lynn has the benefit of varied experience in the area of animal handling. She runs the risk, therefore, of simply listing her many credentials and missing out on the opportunity to present herself—her motivations and concerns. However, she does a fairly good job responding to items in the job description and presenting her enthusiastic and caring personality. Notice the third paragraph, where she describes her reaction to a problem she saw in the waiting room. Although she does not mention the term "self-motivation," she gives an example of that term in action when she talks about the seminars she developed to teach children about animal care. What better information could she have listed as a bullet under her work for Dr. Morris, rather than the obvious "updated pet records?" One would assume any veterinary receptionist would have to do that.

Gregory Benjamin

Job Advertisement:

FIELD SUPERVISOR: 2–3 years experience with roofing and remodeling required. Ability to speak Spanish a plus. Send résumé and references to: Robbyns' Restoration and Roofing, 197 Essex Street, Hillside, DC 98888. EOE.

Gregory does a nice job balancing his construction experience with his academic experience in his cover letter. Although he has only worked for one company, he stresses his varied work experience and ties it in nicely with his Public Health major. He forestalls any concerns that the owner of Robbyns' Restoration and Roofing might have about his being able to handle both school and work by showing that he has already been doing that very well, and gaining valuable time management skills. He even emphasizes that his organizational skills will help the company stay on schedule and improve customer relations. He presents himself as a superior candidate; that might seem brash, but the details he provides back up his claim. What do you think about the tone of the letter? How might his confidence appeal to or annoy his potential employer? When would this tone be appropriate, and when might it backfire?

Steven S. Adamo

Job Advertisement:

SYSTEMS ENGINEER: We are seeking a diligent and energetic person to develop models to improve manufacturing health-related products. Experience in health technologies and working knowledge of relevant computer software is required. In addition, we are looking for someone with a familiarity with team development and management.

Steven's résumé and cover letter show how a student can tailor coursework to meet an employer's needs. What abilities can we assume the employer is looking for in a job candidate? How does Steven present his coursework as work experience? How might Steven express himself more concisely? In the third paragraph, how do the skills Steven describes give further evidence that he is a good worker? How might Steven shift this paragraph from "telling" the reader he is a problem solver to "showing" the reader he is a problem solver?

Candice Common

Job Advertisement:

MANAGER OF COSMETIC LINE: We are looking for a motivated individual to promote our new line of cosmetic products. Qualified individuals will have a significant amount of experience in sales and/or marketing of beauty products. Evidence of successful implementation of innovative business practices is required.

Notice how Candice's cover letter effectively expands on her résumé, describing key details of her employment history. How does Candice frame these details around a potential employer's interests in the second and third paragraphs? For example, why might an employer be especially interested in the "strong clientele" Candice can build? How does Candice preview these key ideas in her introduction? What transitional phases show the logic between Candice's ideas? Examine how Candice uses these phrases to link the chronology of her employment history and to link her cosmetic abilities to her business skills. What do you think about the letter's tone? Are you convinced that Candice has the abilities she lists in her cover letter and résumé? Why?

A Problem of Coherence

How could the central paragraph of the letter be made more coherent, so that it is not merely additive ("In addition," "Also," "And") but instead makes a coherent statement about what the candidate offers?

11111 RPO Way
New Brunswick, NJ 08901

July 21, 2009

Ms. Sara Creighton
Financial Representative
Northwestern Mutual
New Brunswick, NJ 08901

Dear Sara Creighton,

After briefly speaking with you at the Rutgers Career Fair, I am now interested in an intern position at Northwestern Mutual following the spring semester of 2009. I have carefully read through the information packets that you have given to me, and I think I have the skills that your firm seeks.

Employment as a customer service representative for a commercial bank has allowed me to learn the importance of being a team player. In addition, I've worked at Bloomburg where I had the opportunity to work within a large company doing projects for four different departments. Also, these projects strengthened my computer skills with Microsoft Word, Access, and Excel. And I'd like to add that various times throughout this internship I went beyond what was expected in my duties in order to make sure that projects would succeed.

I am seeking a position that would offer challenge and continued growth opportunity. My passion for doing the job right and taking on responsibilities and challenges make me a strong candidate for this position. I would also strive to make lasting improvements and to work in whatever capacity necessary to succeed. You can contact me at 555-555-5555 and at JLA99@eden.rutgers.edu.

Sincerely,

Joel Anderson

Enclosure

JOEL ANDERSON

JLA99@eden.rutgers.edu

CAMPUS ADDRESS	PERMANENT ADDRESS
Rutgers University	41 Allen Court
11111 RPO Way	Ramsey, NJ 07013
New Brunswick, NJ 08901	(201) 111-1111
(732) 111-1111	

EDUCATION: Rutgers University, New Brunswick, NJ
B.A. Marketing, May 2010
G.P.A.: 3.7

EXPERIENCE: **USAA Bank**, Ramsey, NJ   *July 2006–Present*
Customer Service Representative
- Participate in weekly meetings.
- Process various financial transactions for customers.
- Resolve customer inquiries and questions about their accounts using CRT System.
- Collect data to create reports in Access and Excel.
- Assist the bank's personal banking representatives and investment specialist to increase sales and profit by promoting the bank's products and services.
- Recognized formally with two appreciation awards.

Bloomburg, New York, NY   *June–August 2005*
Summer Intern
- Provided weekly reports via Access and imported to Excel.
- Presented ways of improving website to managers and implemented those changes by working with other employees and using HTML.
- Researched E-commerce companies on the Internet.

Outback Steakhouse, Edgewater, NJ   *May 2004–Aug. 2004*
Waiter
- Developed good working habits in a fast-paced environment.
- Responsibilities included: making sure weekly schedules were fair and organized, looking after the upkeep of the floor, hosting.

SKILLS: Microsoft Windows, Microsoft Office, Internet Explorer, HTML
Strong interpersonal skills, superior oral and written communication skills, excellent work ethic.

Major Problems of Specificity and Emphasis

Joan's letter is so general that it could apply to anyone. How can the writer use evidence from the résumé to offer specific details about her abilities? Looking at the résumé, what major problems do you see? What parts of the résumé deserve more emphasis and which deserve less? What part ought to be fleshed out to emphasize her experience with real-world projects even in the classroom?

122 George Street
New Brunswick, NJ 08901

February 4, 2009

Mr. James Barra
Air Cruisers
PO Box 180
Belmar, NJ 07719

Dear Mr. Barra,

I am very interested in working as an Industrial Engineer for your company this summer. I learned of the position from the Internship Guide provided by the Career Services at Rutgers University.

I will be completing my third year as an Industrial Engineering student at Rutgers. I have taken and excelled at manufacturing, work design, and other courses that relate directly to this position and which included real world experience with design projects performed for area businesses. These projects have provided me with knowledge and experience in the practical applications of engineering. I would like to use this knowledge and experience to benefit your company.

I have encloses a copy of my résumé which further describes my educational background as well as my work experience. I look forward to hearing from you soon. Thank you for your time and consideration.

Sincerely,

Joan Parker

Joan Parker

CAMPUS ADDRESS Rutgers University 11111 BPO Way New Brunswick, NJ 08901 (732) 777-1111	**PERMANENTADDRESS** 419 College Avenue Pittsburgh, PA 88888 (814) 111-1111

EDUCATION:	**Rutgers University, College of Engineering.** New Brunswick, NJ B.S. Industrial Engineering, expected May 2010 G.P.A.: 3.29
HONORS:	Dean's List, Fall 2007, Spring 2008 through Fall 2009 Edward Bloustein Distinguished Scholar Member, Alpha Pi Mu, Industrial Engineering Honor Society
RELEVANT COURSEWORK:	Work Design and Ergonomics—*Fabric Co. Consulting Project*, Deterministic Models in Operations Research—*Optimization Project*, Manufacturing Processes—*Salt Shaker Design Project*, Engineering Probability—*Bakery Consulting Project*, Design of Mechanical Components—*Duplexer Device Design Project*, Accounting for Engineers, Intermediate Statistical Analysis
EXPERIENCE:	**Byers Engineering Co., Somerset, NJ, July 2008–Present** Administrative Assistant • Aided in the operation of an engineering office. **Six Flags Great Adventure, Jackson, NJ, Summers 2006–2008** Security Officer • Screened incoming park guests. • Responsible for the handling and logging in of confiscated materials. • Maintained appropriate safety and security conditions. • Trained new officers. **Express, East Brunswick, NJ, June 2006-October 2007** Retail Salesperson • Provided customer service. • Assisted with merchandise displays. • Trained new employees.
COMPUTER SKILLS:	BASIC C SIMON Word FORTRAN LINDO Excel Symphony
ACTIVITIES:	Member, Society of Women Engineers.

Stronger Letters and Résumés

Campus P. O. Box 867
Forrest Glen, DC 88888

April 20, 2009

Dr. Gilbert Sullivan
Companion Animal Hospital
824 Poplar Street
Hillside, DC 98888

Dear Dr. Sullivan:

I would like to be considered for the position of veterinary assistant. I learned of this opportunity from your advertisement in the *Halcyon Press* of April 19, 2009. I am currently a third year pre-veterinary major attending Halcyon University, and I would love to take advantage of the wonderful learning experience working at your hospital would bring. I feel that I could contribute to your practice and relate to your patients in a caring and professional way.

I have previously worked with veterinarians for both small and large animals at school and during the summers. I have learned valuable customer relations skills in dealing with pet owners at veterinary clinics, and also at the kennel where I work to provide people with calm, well-behaved dogs. Currently, I am enrolled in a class that allows the students to do tests on a group of twelve foals in addition to training them. I have taken blood from these foals for glucose tolerance tests and assisted in ultra-sounding them for body fat. Along with my work with Drs. Morris and Bell, I have also visited the Pennsylvania Equine Center and spent the day with one of their vets, Dr. Josiah Brennan. I assisted in a castration and helped him with the various patients he had that day. In the past few years I have worked with a wide variety of animals including horses, dairy cows, dogs, cats, and sheep. Simply, I enjoy helping all animals in any way I can.

At the last veterinary clinic where I worked, I noticed that the children in the waiting room often did not know how to behave around other people's pets—or even their own! With the support of Dr. Morris, I developed a series of seminars to train children of pet owners how to care for and be affectionate with their animal companions. I also taught them how to behave around a strange animal. These seminars were well-received by the children, their parents, and the clinic staff. I am enclosing a copy of my résumé, which provides more information about my motivation and my ability to work well with people and their animal friends. I can be reached at 888-009-4573 or at lynncato@cl.com. I look forward to hearing from you. Thank you for your time and consideration.

Sincerely,

Lynn H. Cato

Lynn H. Cato
lynncato@cl.com

Present
Halcyon University
Campus PO Box 867
Forrest Glen, DC 88888
(888) 009-4573

Permanent
298 S. 18th Ave.
Catesville, PA 10324
(764)-000-9087

EDUCATION	Halcyon University, Forrest Glen, DC Pre-Veterinary Major, Equine Science Minor, Expected date of graduation: May 2010. GPA: 3.2
HONORS	Halcyon Distinguished Scholar, 2008, 2009
RELEVANT COURSES	Animal Science, Animal Nutrition, Animal Reproduction, Dairy Cow & Horse Handling, Horse Management, Animal Problems

EXPERIENCE

Kennel Manager/Dog Trainer, Summer 2008–Present
Noah's Ark, Brattington, VA
- Responsible for care and exercising of all boarded dogs
- Ran dog obedience classes
- Aided visiting veterinarians and groomers

Volunteer/Assistant, Summer 2007
Dr. Maria Bell, Equine Veterinarian, Owens, VA
- Performed flex tests on horses to check for lameness
- Assisted in ultrasounding mares to check for pregnancy

Barn Manager/Horse Trainer, Summer 2006
Turnaround Farm, Catesville, PA
- Responsible for feeding and exercising of about forty horses
- Trained various horses for hunter/jumper competitions
- Taught beginner lessons
- Assisted visiting veterinarians and farriers
- Groomed at various horse events and shows
- Helped customers find an appropriate mount

Veterinary Assistant/Receptionist, 2005–2006
Dr. Bernard A. Morris, Catesville, PA
- Assisted during small animal surgeries
- Responsible for feeding and bathing dogs and cats
- Wrote letters to clients and updated pet records

SKILLS   Microsoft Word, Microsoft Excel, Access, Internet

ACTIVITIES   **Halcyon Student Safety-Mounted Patrol, Fall 2008–Present**
• Patrol campus on horses and report problems to police

Member, Animal Science Club, Fall 2008–Present

Member, Catesville Equestrian Team, 2004–2008

Halcyon University
Campus P. O. Box 984
Forrest Glen, DC 88888
February 5, 2009

Mr. Howard Robbyns
Robbyns' Restoration and Roofing Co.
197 Essex Street
Hillside, DC 98888

Dear Mr. Robbyns:

I am writing to express my interest in the position of field supervisor for your company. I learned of the opening from the February 4, 2009, *Washington Post* classifieds and feel that I am a superior candidate for this job. I can offer you my knowledge of onsite safety and my hand-on experience in construction, as well as my ability to work well with a variety of people.

For several years I have been a supervisor for Herb Chundley Construction, a company handling the remodeling and renovations of homes and modular units. I have worked on several large projects as well as smaller scaled jobs and have dealt with the completion of several projects at the same time. The leadership qualities I have learned over the years will be invaluable in a supervisory position. Both on the job and as a member of Labor Union #934, I have learned how to deal with a diverse group of people at many levels: labor, management, and customer. In addition to having a working knowledge of Spanish, I am fluent in Polish. Most of the projects I have worked on involved either remodeling or installing and repairing different kinds of roofing, both shingle and tar. Being quite familiar with the tools and machinery needed for tasks at hand, I will be able to inform and assist workers with any difficulties they may encounter.

My on the job experiences have led me to a firm commitment to providing a safe working environment, something which my coursework in Public Health and Environmental Science at Halcyon University has reinforced. My experience and training will aid in the prevention of job-site accidents. In order to work fulltime and take a fulltime courseload, I have had to develop my organizational and scheduling skills, something which will further help me complete all site-work on time. As I have learned, staying close to or ahead of schedule is one of the best ways we can promote customer satisfaction and increase repeat business.

I have enclosed a copy of my résumé that shows my work and academic experience. Please contact me at handygreg@cl.com to set up an interview. Larry Anthony of Local # 934 (888-009-5893) and Herb Chundley (888-009-7032) have offered to provide references for me. Feel free to call them. I look forward to hearing from you.

Sincerely,

Gregory J. Benjamin

GREGORY J. BENJAMIN
handygreg@cl.com | (888) 009-4708
Halcyon University, Campus P. O. Box 983, Forrest Glen, DC 88888

OBJECTIVE: Seeking a supervisory position in the field of construction with emphasis on roofing and exterior remodeling.

ABILITIES:

- Skilled at organizing projects and estimating time of completion.
- Speak Spanish and Polish.
- Knowledge of all types of excavation and remodeling techniques.
- Experience in organizing productive crews for job sites.
- Skill with all types of saws and tools needed for repair and remodeling
- Proficient knowledge of Microsoft Excel and Microsoft Access.
- Experience in customer relations.
- Certified in CPR and Lifesaving since 2006.

EXPERIENCE: **Member of Labor Union #934, District of Columbia**

Herb Chundley Construction: *Fall 2008–Present*
- Remodeled homes and modular units, including the interior, exterior, and roofing.
- Organized separate projects according to time, amount of labor, and cost of materials and labor.
- Developed time-saving techniques in a safe working environment.
- Organized extremely large project including the renovation of five multifamily homes.
- Organized large projects at multiple mobile home complexes.
- Supervisor of several large projects dealing with roofing and siding on several different style homes and other types of buildings.

EDUCATION: **Halcyon University, Forrest Glen, DC**   *Fall 2007–Present*
Public Health Major. GPA: 3.5 Major GPA: 3.92
- Courses in Environmental Science and Public Health
- Courses dealing with environmental hazards and precautions and also public safety and awareness.
- Attended several work safety seminars including jobsites located at DC Civic Center, Capitol Hill Complex, and Metro Recycling.

21 Cedarville Ave.
Piscataway, NJ 08940

May 7, 2009

Newton Manufacturing, Inc.
178 Technology Plaza
Fairfield, NJ 65746

Dear Director of Human Resources:

A posting in *The Star Ledger* on March 3, 2009 attracted my attention. I saw that your company desires to hire a systems engineer to develop models for more efficient manufacturing of products for the health industry. I believe my education and work experiences make me a strong candidate for the position, and I hope you will concur.

Later this month I will graduate from Rutgers University, School of Engineering with a B.S. degree in Industrial and Systems Engineering. During my four years at Rutgers, I have completed the full range of courses needed to begin a career. For your needs, two courses have special relevance. Production Control was a course where I learned that perfection is essential when the manufactured product is intended to remedy life-threatening illnesses. Simulation Modeling was another course of significant importance. In that class and lab, we developed prototypes to prepare for actual manufacturing.

In conjunction with my course work, I have been working part-time for Audio Hearing Instruments for the past two years. The tasks I have been responsible for have given me practical experience in solving problems relating to worker efficiency and attention to details.

I hope this letter and the attached résumé persuade you that I am a strong candidate for the systems engineer position. An opportunity to meet with you for an interview for the position would be greatly appreciated. You can contact me by phone at (732) 555-1234 or by email: sadam@eden.rutgers.edu. Thank you for you consideration.

Sincerely,

Steven S. Adamo

Steven S. Adamo
21 Cedarville Ave.
Piscataway, NJ 08940
(732) 555–1234
sadam@eden.rutgers.edu

Objective	To obtain an entry-level position requiring strong analytical and organizational skills in a manufacturing engineering environment.
Education	*Rutgers University, School of Engineering* B.S., Industrial and Systems Engineering, May 2009 GPA 3.65

Relevant Coursework

Manufacturing Processes	Production Control
Design of Engineering Systems	Manufacturing Facility Layout
Engineering Probability	Computer Controlled Manufacturing
Simulation Modeling	Quality Control/Design of Experiments

Computer Skills

Software:	Microsoft Word, Excel, Access, Power Point Micrografx, Stat-grahix, Arena/SIMAN, TK Solver
Languages:	Fortran, C++

Experience

Dec. 2007 to Present

Audio Hearing Instruments, Manufacturing Support Group, Manville, NJ
Industrial and Systems Engineer Intern
• Perform time studies for floor-plan design and assembly line layouts
• Designed fixture/tools for process improvement
• Manage process database
• Provide support to international facilities

Sept. 2006 to Dec. 2006

International Rope Connector
Senior Project
• Group leader of four person engineering team
• Design/constructed/tested universal connector
• Determined facility layout and process for mass production

May 2005 to Sept. 2006

Williams Lumber Company, Sussex, NJ
Truck Driver
• Delivered lumber and other building supplies to construction sites
• Developed communication skills

Awards and Affiliations

N.J. Science Scholar Award 2007
School of Engineering Scholar 2006
Association of Industrial Engineers

123 Castlepoint Terrace
Hoboken, NJ 07030

March 19, 2009

P.O. Box 221
New York, New York 10002

Re: Job ID#23344

To whom it may concern,

I saw your ad at Monster.com, and I noticed that you are requiring a responsible yet out-going individual to oversee your new cosmetic line in New York. Launching a new line requires an individual who is capable of dealing with the multiple responsibilities, such as make-up artistry, a willingness to learn new information, mathematical knowledge for placing orders and budgeting, managerial experience, sales experience, and sociability to help build the client base. I feel that I have the skills and experience that will be required of launching your line in New York.

Currently the counter manager for "Poppy," a budding Australian cosmetics line at Barneys, I have seen the initial difficulties that can come with launching an international cosmetic line, but I also know of the great satisfaction that eventually comes with overseeing the rise of success. With "Poppy" I have not only-built a strong clientele for the line by proving my abilities with the customers, but also with being personable. I am not only their make-up artist providing a service but also a friend. Having worked with cosmetics for a year and a half-now, I have gone from being a novice to an expert in the field - from working as a sales person, to working on movie sets, photo shoots and weddings. As in my résumé, my mathematical skills were also put to the test with the two inventories I had to execute, as well as maintain an accurate stock book. I also assisted with the launch of another new cosmetics line at Barneys- "Cargo." As their resident make-up artist, I coupled my enthusiasm with my artistic ability to boost sales by 45%, as well as create a niche for a once unknown cosmetic line in New York.

My skills in cosmetics as well as business are a unique combination that would prove useful in launching and overseeing your line in New York. I would like to have the opportunity of interviewing in person, so that I may show you a photo-portfolio of my work. I will contact you tomorrow to set up a date at your convenience. Thank you for your time and consideration.

Sincerely,

Candice Common

Objective	Seeking a fulfilling account executive position for a budding cosmetic company.
Experience	*February 2007–Current* <u>Poppy Cosmetics Inc.</u> New York, NY Poppy Counter Manager

- Maintained a cosmetic counter at Barneys NY.
- Acted as a liaison between store buyers and cosmetic vendors.
- Established a strong niche for the new line by building a strong clientele list.
- Produced special events to promote sales and knowledge of the line throughout the department as well as with clients.
- Organized, prepared and executed the last two inventories.
- Responsible for a 30% increase in projected sales through makeovers.

October 2004–February 2007 <u>Cargo Cosmetics Inc.</u> New York, NY
Resident Make-up Artist

- Responsible for the professional application of make-up for all of Cargo's clientele at Barneys NY.
- Increased sales by over 45%.
- Built an extensive clientele list.

July 2002–October 2004 <u>Estee Lauder</u> Seattle, WA
Sales Associate

- Consistently ranked first as part of the sales staff at Nordstrom's.
- Assisted in cosmetic training and promotions throughout the department.
- Worked well in a team atmosphere.

March 2002–May 2002 <u>Warranty and Financial Products Inc.</u> Seattle, WA
Customer Satisfaction and Verification Officer

- Maintained customer satisfaction for this budding corporation.
- Ensured quality content of all sales from the sales staff through verification and data entry of customer information.
- Assisted in the training of 15 new individuals weekly.

Education
Fall 2006–Present <u>Rutgers University</u> *New Brunswick, NJ*
- Currently a pre-business majors.

Fall 2005–Spring 2005 <u>Southern Methodist University</u> Dallas, TX

Reader: _____   **Recipient:** _____

■ Peer Feedback: Cover Letter and Résumé

Please fill out the following form for your partner. Feel free to write comments on the drafts as well.

Does the cover letter . . .
1. directly address the employer? _____ yes _____ no
2. respond to a specific, published job posting? _____ yes _____ no
3. explain why the job candidate is best suited to **this job?** _____ yes _____ no
4. include a high level of detail concerning the strengths
   of the job candidate? _____ yes _____ no
5. appear in full block form and include all six elements
   (return address, date, recipient's address, salutation, body, closing)? _____ yes _____ no

Is the cover letter . . .
1. signed? _____ yes _____ no
2. free of all grammatical and typographical errors? _____ yes _____ no
3. no more than one page in length, in 12 point
   Times New Roman font with one-inch margins? _____ yes _____ no

Does the résumé . . .
1. catch the attention of the reader? _____ yes _____ no
2. include specific, active language? _____ yes _____ no
3. list and describe relevant work and/or academic experience?
   list and describe relevant extracurricular _____ yes _____ no
4. interests and/or activities? _____ yes _____ no
5. provide appropriate contact information? _____ yes _____ no

Is the résumé . . .
1. visually appealing and appropriately formatted? _____ yes _____ no
2. free of all grammatical and typographical errors? _____ yes _____ no
3. no more than one page in length, in a professional font
   size and style? _____ yes _____ no

What parts of the drafts did you like the most?

What parts of the drafts need the most improvement?

Additional Comments/Suggestions:

Researching Your Topic

Research work is like any other work students encounter: a little basic knowledge makes the process more efficient. There are basically five things students need to know to be successful doing research for this course:

- When to use **primary** and **secondary** sources.
- How to judge among **scholarly**, **professional**, and **popular** publications.
- How to research **patrons**, **problems**, and **paradigms**.
- How to find **books**, **journal articles**, and other library resources.
- The proper way to cite sources according to **APA style**.

These five aspects of research are covered in the paragraphs that follow.

Primary and Secondary Sources

How will you show that your topic is important and needs to be addressed. It will not be enough to rely on an emotional appeal or to expect people to take you at your word. Research will be required to demonstrate the nature and extent of the problem in a logical way. Your instructor may require primary research as well as secondary research, but knowing how and when to use them is important.

Primary Research

Primary research, sometimes called fieldwork, is data that you personally collect about the topic. Experiments, surveys, questionnaires, direct observations with note keeping, and interviews are typical examples of primary research. Data you collect in experiments, observations, and surveys can be presented in charts or graphs to quantify the problem. Questionnaires and interviews can be helpful when opinions are important.

Secondary Research

Even if you do collect your own research, you will need other research to interpret your data for others. That is why it is necessary to look at published sources. Secondary research is the term used to describe the search for published information, which you must take at secondhand. The value of secondary sources depends a lot on their credibility (see below).

For your proposal, you might do both primary and secondary research to introduce the problem, but you must do secondary research for the literature review (or paradigm) that helps interpret the problem and explain your solution. Each proposal stands or falls on the quality of its research, and all need a solid foundation of published and authoritative research to support their claims. Without published sources you will be very hard pressed to develop an explanation for your plan of action.

Scholarly, Professional, and Popular: Evaluating Secondary Sources

If there is one thing that students should learn in college, it is that not all information is equally valid or credible. When evaluating sources, students need to keep in mind the types of sources they are, since that will greatly affect the power they have to persuade the reader. Three terms are key: scholarly, professional, and popular.

Scholarly Sources

Scholarly sources are articles and research studies published in peer-reviewed journals. They show what scholars in a particular discipline are thinking about topics based on their research. In the scholarly journals, you will see that discussions reference accepted concepts and models. These readings can be difficult because the contributors to these journals use specialized vocabulary that someone outside of or fairly new to the discipline may not quickly comprehend. Realizing that these sources are the strongest authorities you will have for your proposal should help you persevere even when the reading is challenging. Scholarly sources are found in college and university libraries. Many journals are now in electronic form and accessible on the web, but many are still only in print form. When you access the Rutgers Libraries, you will see whether articles you need can be downloaded or whether you need to go to the library and photocopy or take notes on the information.

Professional Sources

Newsletters, journals, magazines, and websites that are used by the practitioners of a given profession or discipline are known as professional sources. They include up-to-date information about existing and new products, business applications, and commonplaces of the profession. You might find articles there about successful companies or methods written by respected people working in that field. These sources have some authority and can be excellent places to look for models of success. But because the writers of these publications often do not do research themselves and because they often do not take a critical perspective on their specialty or on companies in their industry (where these writers might be employed), professional sources are not considered quite as authoritative as scholarly sources. These publications can often be found in the Rutgers Libraries or through Internet sources.

Popular Sources

Newspapers, magazines, and websites that are readily available to the public and written to a broad audience are generally called popular sources. While they are the easiest sources to find, they have the least value when authority is being established for a proposal that requests funding. Popular sources can, however, supplement the scholarly and professional sources and show how your topic is of general social interest. Many Internet sources would fall under the category of popular.

Based on this brief discussion of the three types of sources, you can see that often the more easily obtained the information is, the less authority it has. The most authoritative sources are generally written for a specialized audience. Recognize the category of the sources you use so you can judge how well they bolster your own authority. Each proposal stands or falls on the quality of its research, and all need a solid foundation of published and authoritative studies, theoretical works, and other documents.

Researching the Patron, Problem, and Paradigm

Often when students begin their research, they see their job as finding out as much as they can about the problem that they want to address. While this can be a good way to start your research, you need to recognize that finding information about the problem is only part of your task. You will also have to do research on funding sources (the patron) and ways of solving the problem (the paradigm). Each part of the project will require different types of research.

Patron

How will you find a funding source? And how will you pitch your project to him or her? You will have to do research to find the best patron for your project and to learn more about what interests them. Often this research is not directly cited in your paper, but it is among the most important in making your paper realistic.

Even if the organization that will be funding your project is the company you currently work for or the school you attend, you will still want to do some research to find out how your project fits with their mission and values. Look at your company website. Look at what is online about your school or about the specific department in your school you are going to ask for funding. How can you connect your project with the issues and problems that concern them?

If there are no local sources of funding for your project, you will need to do some research to see what organizations (including government agencies, private philanthropies, and corporations) share your interests. Here are three good methods for getting started finding a funding source:

Method 1: Go to the Library

The Rutgers Libraries has a wealth of print sources that can help you find funding. These sources are often more complete than sources you find on the web, though they might not be as current or quick to browse. Ask the reference librarians for help getting started.

Method 2: Check Out Online Clearinghouses

There are a number of grant clearinghouse websites, where you can quickly access many groups that provide funding for projects. Some good websites to start your search for funding include the following:

The Foundation Center
http://foundationcenter.org/ and http://foundationcenter.org/findfunders/
This is the best clearinghouse for charity and private philanthropy information.

Catalogue of Federal Domestic Assistance
http://www.cfda.gov/
The official government clearinghouse for all sorts of funds.

Grants.gov
http://www.grants.gov/
A clearinghouse for different granting agencies of the U.S. government.

Community of Science
http://www.cos.com/
A clearinghouse for science-related projects.

National Science Foundation
http://www.nsf.gov/funding/
The NSF sponsors theoretical research in the sciences.

Environmental Protection Agency
http://www.epa.gov/epahome/grants.htm
The EPA sponsors environmental projects.

National Institutes of Health
http://grants1.nih.gov/grants/oer.htm
The NIH sponsors health and health education grants.

U.S. Department of Education
http://www.ed.gov/fund/landing.jhtml?src=ln

Method 3: Browse the Web

Since most organizations who might fund your project probably have a website or are listed on the web, a search engine, such as Google, is not a bad initial search tool. Try entering your keywords for your topic, perhaps along with the words "grants" or "funding," and you should at least get some hints about who is interested in your subject area. This method involves a lot of trial and error, and you are better off starting with Method 1 and Method 2. But doing a general web search should at least give you a better sense of your topic and who is interested in it.

Problem

How can you prove that there is a problem? And how can you emphasize its importance? To make a good case, you will have to do some research on your topic with the goal of finding numbers or of defining your problem well enough to understand its scope.

Before you begin your research on the problem, it's a good idea to think about the specific information that would be useful to your case. Some questions to consider:

- What are the most important numbers needed to convince your patron that this problem is important to address? How can you quantify its scope and scale?

- Can you conduct some of this research yourself, or use research that you have already done? Or will you need to rely on secondary sources of research?

- In order to quantify the problem, what are your best sources of documented evidence? What secondary sources might have information that can help your case?

- Which groups or organizations might have already studied the problem? And where might they publish their findings?

If you can get good numbers, you will be able to make especially powerful visual aids.

When You Can't Find the Numbers You Need

- *Keep trying.* Often, especially with online research, key information is hidden behind the keywords that you haven't tried. For instance, say you are writing a project on making a community service project mandatory at a local high school. You need to find information about teens and community service or volunteering. A search in Statistical Universe using the keywords "teens" and "community service" or "volunteerism" will get you nowhere. The perfect graph for your project can only be found under "surveys—opinions and attitudes, by age." Start early and be persistent. Don't do your research when you are pressed for time.

- *Try extrapolating.* Often it is possible to take percentages from national studies and use them to make educated guesses as to how many people will be affected by an issue on a local level. In order for this to work well, your local population must be entirely typical with the rest of the larger area. For instance, if you absolutely can't find rates of smoking for your town, you could use state or local averages and then work out the equation. If 30 percent of people in New Jersey smoke, one could assume that 30 percent of people in Paterson smoke. However, if your local area is different in some significant way from the larger population, you should not rely on extrapolation.

    A town populated by a significant number of young families cannot be compared to a town with several senior citizen retirement villages. If you are reduced to documenting your local problem by extrapolating from national statistics, you must be honest about it and clearly show how you have arrived at the figures you are using.

- *Fill in the gaps with primary research.* Sometimes a problem is so new or so local that there is not a large amount of hard data to draw from. In that case, you will have to do some surveys or other primary research. As much as possible, make sure that your surveys are legitimate and convincing. Your sample size must be large enough and varied enough to be representative. The fact that twenty of your friends say they dislike Economics 101 is not good evidence that a university should drop the course. As you survey, keep track of what day and time you did the survey, how many people responded, how many of each gender, age, and so on, depending upon the subject of the survey. You should also ask your survey questions in such a way that they will generate good statistical responses. If you conduct your surveys well and present them carefully, they can enhance your credibility. For example, which of the following two statements seems most convincing and why?

    - Fifty percent of the people I surveyed disliked Economics 101.

    - Out of 1,000 students, 50% stated that they ranked Economics 101 (on a scale of 1–10, 1 being the lowest) at 3 or below.

- *Use uncertainty to your advantage.* Sometimes a lack of knowledge is the best evidence you have that a problem exists. Scientists use uncertainty all the time in order to show that more research must be done. If you are writing a research proposal, you should use the lack of statistical information as part of your documentation of the problem. Be sure to discuss the possible dangers or lack of opportunities that result from "not knowing." Perhaps a central part of your project could be to gather data.

Paradigm

How do you support your claim that your plan is the best way to address the problem? A paradigm gives you that support. You might think of it as the research-based rationale for your plan. It authorizes your claims about the problem and justifies your methods.

If you are doing a scientific research project in any given field, your paradigm will derive from previous research. That previous research offers you both examples of practice (what experimental methods did they use?) and a way of understanding the results (how did the experiment support the hypothesis based on previous theory?) Defining your paradigm outside of the hard sciences is not as straightforward, because there is usually not as strong a consensus as there is in the sciences about which methods and theories are best. But you can still use the model of the sciences to guide you in researching support for your plan.

You might think of paradigms as ideally having two parts, along the scientific model: **models of success** and a **theoretical frame**. A model of success is an example of how others have successfully addressed the problem in some other context. A theoretical frame is a language for explaining how a certain solution will work.

Let's say that you wanted to take on the problem of crime on campus. To find research to justify a plan of action, you would want to search for models of success and find a theoretical frame. You may not need both of these things to succeed in writing a persuasive proposal, but looking for them will help you in many ways.

Searching for a Theoretical Frame

If you were studying sociology or law enforcement, it would be logical for you to take on the problem of campus crime since your previous studies had already prepared you for the issue. You might already have an idea, in fact, of what theoretical frames might relate to the issue. If you don't, then at least you would know where to look to find out. You could talk to professors. You could look in your textbooks (especially in their bibliographies). Ultimately, though, you will need to do some research to see what others have written in journal articles and in books about ways of addressing the problem. That's where you will find your theoretical frame and the language you will need to explain it.

To address the problem of campus crime, you would want to look at what researchers have written in the areas of sociology or law enforcement (the fields that seem most applicable to your problem—though other fields might offer ideas as well). One theory of crime you might encounter in your reading is the "broken windows theory," which suggests that if you address small crimes (such as broken windows) you will be addressing the larger issue because, for one thing, small crimes and big crimes are committed by the same group of offenders.

Searching for Models of Success

If you were looking to address the problem of crime on campus, you would logically look at programs at other schools that helped to reduce crime. You would probably also want to look at towns and cities, since they are also potentially good models. You could look for these models in a number of ways.

- You might look in professional or popular sources, such as college journals or newspapers and magazines, which might have stories about successful crime stopping initiatives.

- You could look on the web, where schools might have posted information about their programs (especially those that proved successful).

- You might ask experts in law enforcement who they look to for models, and then try to interview people involved with the programs they suggest.

- If you have already begun your theoretical research, you may have come across some examples in scholarly sources that you can then try to find out more about.

Once you found those models of success, you would want to repeat the research process to see what specific information you could turn up about how and why those programs worked. The more information you could turn up the better, since you will need that research to justify your own choices in constructing a plan.

Merging Theory and Practice

To construct a coherent project, you will need to merge theory and practice so that your theoretical frame explains the model of success you are using to justify your plan. A good example of this merging of theory and practice in the case of campus crime would be to use the model of the New York Police Department who made their city the safest in the nation by cracking down on low-level street crime (from petty theft to vandalism), following the logic of the broken windows theory. You could use the NYPD as your model and draw examples of good practice from them and explain them using the language of theory.

White Paper Assignment

A white paper is a document which describes a current problem. Your white paper will help you begin documenting and quantifying an actual problem in anticipation of the midterm letter. In addition, you will have the opportunity to present information in light of the needs of your chosen funding source.

The white paper should help you collate and organize information and test the viability of your topic. Pay close attention to the **scope** of your potential project. This is the time when you should be aware of your ability to fully address the problem identified. Upon further review, if the problem still requires additional narrowing or framing, this is the time to consider the possibilities. Your white paper should be brief (one to one and a half pages) and include a significant amount of **fieldwork**. When drafting your white paper, consider whether a possible proposal does all of the following:

1. **Identifies with people**: Does the writer have a particular reader (or funding source) in mind? Does the writer's approach seem appropriate to the reader's concerns? Should the writer imagine a different reader for the idea or find out more about the reader's concerns? Does the project address the needs of a particular population? Might the interests of the reader differ from those of the population to be served? How so?

2. **Points to a problem**: Does the writer demonstrate a need for this proposal? Has he or she discussed a problem that could be researched and documented? How might the writer find out more about the problem? What sources of information might be helpful? What types of evidence would help illustrate the problem better?

3. **Faces complexity**: Is the idea of sufficient complexity to require a detailed proposal? If not, can you suggest ways to develop the project so that it would be adequately complex? Has the writer considered all the major problems here, or is there something he or she is avoiding?

4. **Suggests lines of research**: Does the topic lend itself to library research (a course requirement)? What other kinds of research should the writer consider? How might the writer support his or her claims about the problem suggested by the proposal?

5. **Positions the work within a paradigm**: Does the writer have a definite approach to the problem or issue? How might the writer position him or herself within a discipline or field of study in approaching the topic? What disciplines might be helpful? What research might the writer pursue in developing the paradigm?

6. **Demonstrates originality**: Is the specific work proposed at least somewhat original? Has this idea been tried before? What could make this idea more innovative? Are there other ways of approaching the problem?

7. **Stays within reach**: Is the proposed idea manageable? In other words, is the scope of the proposed work something that can be done well, given the time frame and resources? Is the student remaining within his/her reach, if not his/her grasp? Is the idea focused enough in terms of population, location, or issue? Is it something that could actually get done? Can you see this student actually taking on such a project now or being able to do so within the next few years?

Finding Books, Journal Articles, and Other Sources at the Library

Today there can never be the excuse that you "couldn't find any research" on something. You will see, in fact, that there is usually too much information on any topic. Just try a search on the Index called "ABI / Inform" (from the Rutgers Libraries home page, http://www.libraries.rutgers.edu/, click on "Indexes and Databases" and click on "ABI / Inform" which is listed near the top in alphabetical order). Enter keywords about your topic and you should find that there is lot of information out there (much of which is accessible online in full-text format). And if you try a search at Google, it is likely you will get too many hits to look through in a sitting. You must learn to be selective, have confidence in your ability to analyze what you read, and just simply get to work.

The best place to start is the Rutgers Libraries home page, http://www.libraries.rutgers.edu. If you are using a computer off campus, it can be set up to access the Rutgers Library: on the library's home page, click "Off-Campus Support" for instructions. The library home page is set up for easy use, listing the "IRIS Catalog" (listing all book and periodical holdings at all Rutgers Libraries), "Indexes and Databases" (offering expansive bibliographic references, and sometimes full text, of articles in various fields), and "Subject Research Guides" (offering links to library resources and quality websites reviewed by Rutgers librarians).

The only way to learn how to use the library or its home page is by using it. But if you have trouble getting started, there are tutorials online (click "Learning Tools"). Your class will also have a library tour to familiarize you with the resources available at Rutgers, so be prepared to ask questions of the reference librarians. Remember the reference librarians can be the best teachers of library skills; the library is their classroom, and you are their students. Show them what you have done; ask them questions; seek their advice whenever you get stuck looking for information. The more specific your question, the better the help you will receive.

If you can't locate a source at Rutgers, you can order any book or journal article through interlibrary loan, usually very quickly (no more than two weeks). If you start your research early, you should be able to get all the information you need. Be careful, however, to continue your research efforts while awaiting sources you have ordered. The deadline for completing an assignment will not change if your ordered source does not arrive on time or proves less than helpful.

Some Advice on Searching the Internet

You should never rely upon general Internet searching as your main source of information. Internet sources tend to be too simplified and too much driven by self-interest to serve as the basis for your research. You should always seek a wide variety of sources, using books for depth of coverage, peer-reviewed journals for thinking in your field, and periodicals for timely coverage of recent events. The Internet should be only a supplement to these sources. These suggestions are therefore intended to give you some ideas about using the Internet as an assistant rather than a crutch.

- Often web searches can help you most in developing a list of keywords that you can use later in searching through databases and books. Try putting quotes around specific phrases, like "binge drinking" rather than binge AND drinking, since this will help narrow your search to only sources that use those words together. Remember the basics of Boolean logic: use "and" to narrow and "or" to expand categories.

- If you are beginning with a broad subject (cancer, AIDS, alcohol, guns), try starting with the "web guides" prepared by Rutgers Librarians (accessible from the Rutgers Libraries home page at http://www.libraries.rutgers.edu). Use the guides to help narrow your topic and to get ideas.

- An increasing number of statistical sources are available online. You can also use *Statistical Abstracts of the United States* and *Statistical Reference Index,* which are available in the reference section of most campus libraries. If you are seeking government statistics, check out "thomas," the government center for information at http://www.thomas.gov/. For New Jersey information, try http://www.state.nj.us/. For census information, go to http://www.census.gov/. If you were looking for statistics on campus crime (as in our example above), you would definitely need to visit the Office of Postsecondary Education's Campus Security Statistics Website at http://ope.ed.gov/security/.

- If you find a good website, see if it contains links to others or lists of references you can find in the library. Often, web sources are abstracted versions of much better journal articles or books. Go to the original source!

A Brief Guide to Using APA Style

The following guidelines are not intended to be all-inclusive but merely to help you avoid typical pitfalls in citation. For the purposes of this class, you should use citation style as given by the American Psychological Association. You will need to know APA style for both in-text citation and your References page. The following are guidelines based on current APA recommendations. For more information on the intricacies of APA citation, consult the *Publication Manual of the American Psychological Association* (available in the reference section of all campus libraries). For the latest recommendations on electronic references, go to the frequently updated APA website at <http://www.apastyle .org/elecref.html>. The following examples of format are all based on those sources.

In-Text, Parenthetical Citation

The main purpose of in-text citation is to link information in your text with entries on your References page. For that reason, you need to make the connection between citations and sources clear by using the same primary name in your text as the primary identifying reference in your References. Because science is rapidly evolving, APA citation emphasizes the date of publication by placing it second to the author's name. Therefore, when using APA style the two most important pieces of information you should have in a textual reference are the last name of the author(s) and the date. The APA suggests that you include a page number (or paragraph number if no page number exists, as on a web source) only if you are using a direct quotation from the text. If you mention the author(s) in your sentence, then you should only put the date in your parenthetical citation. Three examples:

> According to J. Q. Wilson and G. L. Kelling (1982), "at the community level, disorder and crime are usually inextricably linked, in a kind of developmental sequence" (p. 33).

> According to a classic study of the "broken windows" theory, "at the community level, disorder and crime are usually inextricably linked, in a kind of developmental sequence" (Wilson & Kelling, 1982, p. 33).

> According to the classic study of the "broken windows" phenomenon, disorder leads inevitably to crime (Wilson & Kelling, 1982).

In parenthetical or in-text citation with three to five authors, you should include all of the names on the first reference; then from the second reference on, use the primary author's name, followed by "et al." With six or more authors you should cite the primary author and indicate others with "et al." For example:

One author:	(Jordan, 2001)
Three authors:	(Jordan, Slinkoff, & Presser, 2001)
Six or more:	(Jordan et al., 2001).

If the source has no discernable author, then use the title. And be sure to use the title for reference both parenthetically and in your works cited:

> *Parenthetical citation for unpaginated, non-authored source:*
> Safir's first action was to focus on the seemingly "trivial" crime of jumping subway turn-stiles to avoid paying the fare (Commissioner describes NYPD 'success story,' 2000).

> *References listing for unpaginated, non-authored web source:*
> Commisioner describes NYPD 'success story.' (2000, January 28). *Yale Bulletin and Calendar, 28*(18). Retrieved March 3, 2009, from http://www.yale.edu/opa/v28.n18/story4.html

Non-Accessible Sources

Since the whole purpose of including citations and references is to provide your reader with the means of finding the same sources themselves for future research, you should not include the following as part of your bibliography: e-mail, personal interviews, phone communications, references to non-published lectures or speeches, and surveys or other original research you have done gathering information about the problem you are addressing in your proposal. You should, however, mention them in your text. Original research that you have done to find information about your project should especially be explained clearly. Here are some examples of how you might cite this material:

> A survey of 45 Busch campus students conducted at the Busch Student Center on April 1, 2003 (see Appendix I for questionnaire and results), showed an overwhelming number avoided taking Friday classes.

> In an interview on January 12, 2003, Robert Spears, the Director of Parking and Transportation for the Rutgers, New Brunswick campus, discussed some of the problems that made additional parking spaces on College Avenue Campus impractical.

> In a March 10, 2003 e-mail response to my inquiries, Professor Dowling said that he thought student evaluations "put pressure on faculty to do the popular thing rather than the right thing" and therefore ought to be replaced by another system.

Your References Page

A References page is just that: it reflects the works that you have actually referenced in your text, not works that you consulted for background information but did not cite. According to the latest guidelines, it should be double-spaced consistently throughout. Do not put extra spaces between entries. You must list your sources in alphabetical order, either by the author's name or the title. If you have two or more sources by the same author, list them by year and use the author's name for each. Where the author has two entries for the same year, add lowercase letters ("a," "b," etc.) to the date. Do not number your entries on the page—alphabetical order and indentation will separate one entry from the next.

The following examples will give you some idea of format; for more complete information, consult with your instructor or the *APA Handbook.*

Books

Jacobs, J. (1961). *The death and life of great American cities.* New York: Random House.

Books with more than one author

Wilson, J. Q. & Herrnstein, R. (1985). *Crime and human nature: The definitive study of the causes of crime.* New York: Simon & Schuster.

Book chapters

Wilson, J. Q. & Kelling, G. Broken windows: The police and neighborhood safety. In J. Q. Wilson (Ed.), *Thinking about crime* (pp. 77–90). New York: Vintage Books.

Periodical articles

If the periodical is paginated continuously throughout the year, only the volume number is needed. If each issue begins with page 1, include the issue number after the volume number, with the issue number in italics and the volume in parentheses. For example, for "volume 17, issue 4," use *17*(4).

Brown, L. and Wycoff, M.A. (1987). Policing Houston: Reducing fear and improving service. *Crime and Delinquency, 33*(1), 71–89.

Strecher, V. (1991). Revising the histories and futures of policing. *Police Forum, 1,* 1–9.

Newspaper articles

Remember to introduce page numbers with "p." or "pp." in the case of newspapers.

Campbell, G. A. (1997, October 14). Crime is down all over. *New York Times,* p. D14.

Unpublished, in-house documents

Brown, M. (1997). Log of daily accounts. (Available at Bryan Dentistry, Westerly Place, West Park, NY 02984).

Web references

For referring to documents on the web, please check the APA website http://www.apastyle.org/elecsource.html for the latest recommendations. For most web documents, you can use the same format as non-web sources, but you then need to add "Retrieved Month Day, Year, from the website: http://etc.org/whatever.html." Differences arise because of the impermanence of web sources and the fact that many do not have clear authors or titles on their pages. The impermanence of web sources makes it necessary to add the date you retrieved it. In the case where there is no listed author, then use the title as your main listing, and only if there is no author or title should you merely list the source (such as the sponsor of the website). In any case, be sure to list the web address of the actual article or page you are using, not simply the address of the website you accessed first. Do not expect your reader to be able to follow all of the links you followed to find the article. If you are only referring to a website in general, and not a specific article or page on the site, you can just record the name and web address in your text without including any specific listing in your References. Do not include a period after web addresses.

Muzzey, E. H. (2001). Biochemical reactions in toddlers: The effects of television on the lymphatic system. *Journal of Northeastern Medicine, 36,* 90–123. Retrieved April 7, 2009 from the JNM website: http://www.jnm.org/journal/muzzey/23000.html

Sample Annotated Bibliography

An annotated bibliography is simply a preliminary References page to which notes (or "annotations") have been added after each entry. The main information required would be a sentence or two summing up what the source says and how it will be useful to your project. You might also want to say whether you will be using the source to quantify the problem or to set up the research paradigm for your project.

References

Bocamazo, L. M. (1991). Sea Bright to Manasquan, New Jersey: Beach erosion control projects. *Shore and Beach, 59,* 37–42.

This paper shows beach control projects that have been used in New Jersey in the past fifty years. It will help me decide the best method for stopping the erosion at Bayport. Although not all of the projects deal with a similar ecosystem, there may be some that will function as models for my own.

Jackson, N. L. (1996). Stabilization of the shoreline of Raritan Bay, New Jersey. *Estuarine Shores: Evolution, Environments, and Human Alterations, 17,* 397–420.

Seawalls are discussed in this article, along with other methods of alleviating erosion. This will help me decide if a seawall would be helpful at Bayport, and if so, will help me determine what height the seawall should be.

Nordstrom, K. F. (1989). Erosion control strategies for bay and estuarine beaches. *Coastal Management, 17,* 25–35.

A seawall alone may not be enough to control the erosion and sea level problem at Bayport. The seawall combined with other structures and techniques to hold back the sea would be a strategy I might consider, and this paper presents models used along bays like the Tranry Bay. Additional control structures will probably be needed. This source is important to my paradigm.

Thorstein, G. (2000). *Land-sea barrier methods.* New York: Putnam.

This book will help me design a good plan for alleviating the effects of sea-level rise in Bayport. Thorstein's failure analysis of methods that did *not* work well will be useful for me in avoiding unsafe or unsuitable designs. Since he includes many sample projects carried out in areas similar to Bayport, I can use them to develop an approach based on an existing paradigm. This source is therefore very important to my paradigm.

▪ Peer Feedback: Annotated Bibliography

Please fill out the following form for your partner. Feel free to write comments on the draft as well.

1. Is the document clearly labeled as a list of references
    at the top of the page?   _____ yes _____ no

2. Does the document contain a minimum of six sources?   _____ yes _____ no

3. Are there various types of sources represented (books to develop a
    theoretical framework, scholarly journals for detailed models, etc.)?   _____ yes _____ no

4. Are at least 50 percent of the references cited as print sources?   _____ yes _____ no

5. Is the document formatted in proper APA citation style
    (alphabetized, indented after first line, publication elements
    ordered correctly, etc.)?   _____ yes _____ no

6. Is the document correctly spaced, in 12 point
    Times New Roman type, with one-inch margins?   _____ yes _____ no

7. Is each entry annotated and detailed in describing how the
    corresponding source would be useful to the plan?   _____ yes _____ no

8. Is each annotation 100–150 words in length, single-spaced,
    and presented in a clear, readable form?   _____ yes _____ no

9. Do the bibliographic entries suggest a theoretical framework
    for the plan?   _____ yes _____ no

10. Do the bibliographic entries include models of success appropriate
    to the plan?   _____ yes _____ no

11. Based upon the entries, is there evidence of a recognizable
    paradigm (or rationale) for the plan?   _____ yes _____ no

12. Is the document free of errors in grammar, usage
    and/or sentence structure?   _____ yes _____ no

What is the one part of the draft you liked the most?

What is the one part of the draft that needs the most improvement?

Additional Comments/Suggestions:

The Midterm
Sales Letter

The Assignment

Write a four- to five-page business letter or memo, single-spaced, not including the list of references, that accomplishes the following:

- Represents the initial correspondence to your patron
- Addresses a specific person by name
- Explains the current problem
- Explains at least some of your initial research toward a solution (your paradigm)
- Cites your research clearly (according to APA style)
- Gives a sense of your plan of action and associated costs
- Closes with an invitation to your oral presentation
- Appends a bibliography of at least eight sources, cited in APA style (remember, though, that at least ten sources are required for the final proposal)

The midterm sales letter should be written as a **letter of persuasion,** and as such it carries the added burden of addressing a particular reader and using some of the means of persuasion available to you for appealing to him or her (with special attention to rational or logical appeals).

Requirements

The midterm paper will be graded according to how well it does the following:

- Adheres to proper letter or memo format
- Discusses, documents, and tries to quantify the problem
- Highlights the reader's concerns about that topic
- Cites specific facts and examples from your research
- Briefly proposes a plan and provides a rationale for it
- Convinces your reader to hear more

- Provides a bibliography of sources in APA style
- Is proofread for errors and appearance

Purpose

The midterm sales letter serves many purposes:

- As a draft of the final proposal, it provides you an opportunity to organize your research toward a practical goal and to begin presenting your information clearly.

- As an evaluative tool, it allows you to receive feedback on your work thus far, so you can have a sense of where you stand with your proposal and in the class.

- As an exercise in persuasive writing, it gives you practice in the most valuable form of writing for business.

Typical Pitfalls and Problems

Students typically go wrong with the midterm paper in the following ways:

- They do not address a specific person capable of funding the project.
- They fail to provide evidence of the problem or trend they seek to address.
- They fail to explicitly cite their research.
- They assert things without evidence.
- They fail to attach a bibliography of sources.
- They use insufficient or inappropriate sources.
- They are poorly proofread for errors and appearance.

Some General Advice, or "14 Steps to a Strong Sales Letter"

You have already gained some practice in writing the letter of persuasion when you wrote the cover letter with your résumé. Here you are also making a sales pitch, but in a much more detailed way. There are fourteen things you will want to consider as you write it. These are stated as rules, and some are always good rules to follow. Obviously, each situation should dictate the type of approach you take. Also, these ideas should not limit your creativity. Remember that the audience should always direct your approach. Who will read your letter? What are your reader's concerns and interests? How can you appeal to this reader most powerfully? How can you explain your evidence? The answers to these questions should guide the way you write the sales letter, and they will always vary from situation to situation. So accept what follows as friendly advice:

1. **Know your audience.**

   Knowing your audience might require some preliminary primary research, or fieldwork. If you are responding to a specific request for a proposal (commonly called an RFP), then you will know some of what your audience expects. You will usually be addressing someone you do not know very well at all. Find out what you can. What is the corporate culture like at your reader's organization? What is their motto or corporate philosophy? What image do they project in their advertising? What recent endeavors have they undertaken? What problems are they facing? What is their competition up to? Find out about your reader's general interests so that you can know better how they might fit with your idea. What specific benefits can the individual or organization you plan to address gain from solving the problem or responding to the trend you are considering?

2. **Get the right name, and get the name right.**

Address your letter to a specific person whenever possible—and, for the purposes of this class, ALWAYS. How many times have you seen a letter that opens, "Dear Sir or Madam"? Does that inspire much interest in you? Not only is a letter addressed to a specific person bound to generate a more positive response, it will more certainly be read—and it will more likely be read by that specific person capable of making a decision on your project. (The success of annoying ads like Publisher's Clearinghouse is due in no small way to the appearance of personal interest: even the most cynical readers are unconsciously and unavoidably flattered by the fact that Publisher's Clearinghouse knows their name).

How do you find out the person to whom you should address the letter?

This is another one of those "legwork" things, but fortunately these days it doesn't require any walking around: usually a simple telephone call or a "visit" to the company website is all you need. This is part of the fieldwork, or primary research, discussed in Chapter Four.

When in doubt, just ask! Call up the company and ask a receptionist. Talk to a few people—maybe even speak to the person you plan to address (that will give you a better sense of his or her style and will provide a good introduction to your letter). Just ask, and be nice about it. Who would handle the sort of project you have in mind? What department? What person in that department?

Once you know who you should address, find out how you should address that person. How do you spell his or her name? For purposes of the oral presentation, you will want to know how it is pronounced. Does he or she have a title? Does she prefer Ms., Mrs., or Dr.? Is there a middle initial? A Jr., Sr., or Roman numeral? Find out.

3. **"Dear" is never wrong as an address.**

"Dear" is the expected mode of address. Though you may have struggled in personal correspondence over whether or not to write "Dear" to your reader, in business correspondence it is simply a standard formality.

4. **Make a strong first impression.**

How you open your letter will depend upon the specific audience and the specific appeal you want to make. If you know the addressee, you will likely want to remind him or her of that fact and allude to your most recent or most positive interaction. If you don't know the addressee personally, you'll have to be more creative. You can rarely go wrong by trying to open with a confident and definitive statement, and you should open emphatically whenever possible. Point to the problem or need you seek to address, or state the sort of vision you will provide in responding to this need. Get this person to read further.

5. **Show that you identify with your reader's concerns.**

Explicitly state what you know about your audience's interest in the idea you will propose or the problem you seek to solve. Explain why this person is the most appropriate addressee. Show that you can see things from the reader's perspective, and that you see the proposal as a win-win situation. Your funding source will want to know what is in it for them.

6. **Specify and quantify the problem or need you seek to address.**

If you can quantify the problem, you can show its magnitude and importance. Alternately, you might give an anecdote or example that helps highlight the importance of this problem to your audience.

7. **Get to the point.**

There are some cases where you may wish enigmatically to string your reader along before revealing your specific project. Usually, though, readers in business don't have time to read a mystery novel. So don't keep your reader waiting too long for your discussion of how you intend to solve the problem or respond to the recent trend you have identified. If you offer a deal, be up front about it. What are you offering? What do you want in return? Give your reader a forecast of what to expect.

8. **Provide evidence and examples.**

This is the key to a successful letter for this course. You must cite your research. You must also show that you can use the information you have collected to construct an effective argument for action. You might say that it requires putting information into action. Evidence is always logically persuasive.

9. **Activate your reader's imagination.**

Invite your reader to engage with your idea, perhaps by using rhetorical questions. Get your reader to participate in your text.

10. **Encourage empathy.**

Now that you have shown your reader that you see things from his or her perspective, start to turn the tables a bit. Get the reader to identify with your reasons for being involved in this project, and present your reasons in the best light possible. If your ethos is key to your appeal, you may consider highlighting it earlier in your letter.

11. **Close with a call to specific action and further contact.**

Make sure that the reader sees this as a pressing need, with a deadline for action. For the purposes of your sales letter for the class, you must invite your reader to hear your presentation, listing the specific date, time, and location.

12. **Make contact easy.**

It is always a good idea to provide a way for your reader to contact you easily, either by phone or e-mail. Don't forget to put that down, usually in the last paragraph—especially if it isn't clearly printed on the stationery you use.

13. **Sign off "Sincerely."**

Don't get fancy with the closing address, unless it is especially appropriate to offer "Best wishes." Like "Dear" at the outset, "Sincerely" is the standard close.

14. **Follow up and be persistent.**

Many times you will discover that your letter has languished in the wrong department or that a busy addressee has failed to take any action because the letter has gotten buried under more pressing work. Follow up your letter after a reasonable interval, perhaps with a phone call or another method of contact. Don't give up.

Sample Papers

The sample papers that follow are rather typical of the work that students turn in at midterm. Generally, they are competent samples. However, in line with the chronology of assignments, they all need to be improved to make them more coherent and fully developed projects.

Since the Six P's represent the process of writing the proposal, in order, it is not surprising that most midterm papers do a good job of identifying an appropriate patron to fund the project, defining a population to be served, and trying to understand the problem they want to address, but that they also might be rather vague about their paradigm and the plan. Of course, the price can never be definitive until the plan is sufficiently detailed. Some vagueness is natural, but the better midterm papers will still suggest a more coherent sense of project and will do more not only to describe a paradigm but also to show how that paradigm informs the plan. Since each element of the plan must be justified by published research, you can't possibly have all of the parts of your plan in place until you have identified and integrated all of your sources. However, in the midterm sales letter all of the Six P's should be represented in some way.

The following papers are representative examples of student submissions. They are intended for discussion purposes only and should not necessarily be taken as models of strong work.

Jane Hamilton
Rutgers, The State University of New Jersey
341 CPO Way
New Brunswick, NJ 08901
(908) 538-6678
jham@eden.rutgers.edu

March 16, 2009

Mr. Patrick Spagnoletti
Superintendent of Schools
510 Chestnut Street
Roselle Park, NJ 07204

Re: Reducing the rate of adolescent obesity in Roselle Park

Dear Mr. Spagnoletti:

Obesity is an epidemic that is growing at an increasing rate in today's society. There are various health risks associated with this disease, most of which are terminal. The scary fact of this epidemic is that it is preventable. What is even scarier is just how large the population of adolescents affected by obesity is. Whenever there is a problem there is most certainly a plan to solve it. I have an insightful plan which addresses the problem of adolescent obesity in today's society, starting with children living in the small borough of Roselle Park. Since you have control over the budget, I am hopeful that you will support my conclusion regarding the plan to control adolescent obesity, and give your fiscal support through budget funding to accomplish this project.

The Problem

Adolescent obesity has become a huge issue in today's society. Two studies conducted by the Centers for Disease Control and Prevention in a 1976–1980 survey and then in a 2003–2004 survey show that in children aged 2–5 years, the prevalence of overweight increased from 5.0% to 13.9%; for those aged 6–11 years, prevalence increased from 6.5% to 18.8%; and for those aged 12–19 years, prevalence increased from 5.0% to 17.4% (CDC, 2007). In my opinion these statistics are incredibly alarming. A near three time's increase of something that can be controlled and amended is completely unacceptable. Not only are these statistics alarming, but the correlating health conditions are frightening as well.

There are an incredible number of health conditions which are directly related to obesity. Children face a future laden with hypertension, dyslipidemia, Type 2 diabetes, coronary heart disease, stroke, gallbladder disease, osteoarthritis, sleep apnea and respiratory problems, and even some cancers—endometrial, breast, and colon (CDC, 2007). These negative effects are incredibly detrimental and horrific. It is sad to think of how many of today's youth will suffer from such horrific consequences of their poor and uninformed decisions regarding their current consumption of food and its direct relation to their overall health. Another alarming fact is that overweight adolescents have a 70% chance of becoming overweight or obese adults (Health and Human services, 2007). This statistic proves that poor adolescent habits in relation to nutrition, health, and an active lifestyle carry over into adulthood.

According to the surgeon general, overweight in children and adolescents is generally caused by lack of physical activity, unhealthy eating patterns, or a combination of the two, with genetics and lifestyle both playing important roles in determining a child's weight. Our society has become very sedentary. Television, computer and video games contribute to children's inactive lifestyles and 43% of adolescents watch more than 2 hours of television each day. Children, especially girls, become less active as they move through adolescence (Health and Human services, 2007). So the main problems are rooted in unhealthy choices of food consumption and lack of physical activity.

In an interview I conducted on February 20th, 2009, with Kevin Carroll, Physical Education teacher at the elementary schools in Roselle Park, his insight regarding obesity supported the conclusions of the surgeon's general. Mr. Carroll has been involved with Physical Education for 30 years and has seen trends reflecting an increase in more sedated behavior and poor eating habits. When I asked him his opinion on why obesity is such a huge issue in today's society, he offered a lot of insight with the following:

"I think the diet of kids is to blame. People are busy working and don't have time to cook, its easier to pick something up from McDonalds. It's easier for them to do that, than prepare a meal. Computer games and video games are to blame too. Also, I think people are scared to let kids just play on their own because of all the creeps out there. But I think diet is a big thing. Fast food is relied on a lot, And kids to me have become more sedated in general. They would rather watch TV and play computer games in general."

This information frightening and overwhelming, and I am sure you feel the same concerns that I felt when I came across this startling information. The fact of the matter is that these statistics are not just numbers; they are people, children, with bright and promising futures. The data listed above is on a national level. I am currently working with Mr. Carroll collecting data regarding the population of adolescents in Roselle Park. My plan for Roselle Park will tackle the causes of obesity head on, and help the children of Roselle Park build a strong foundation on which they can make healthy and wise decisions regarding food consumption and activities in their lives.

Plan

As mentioned earlier, the data I collect will help me narrow the broad range of possibilities regarding obesity. However, my plan and course of action will be setting up educational programs at the elementary schools. I will conduct fun, educational activities about health and nutrition. The activities will range from teaching the children why the body needs food to why certain foods are better than others, and how to make good decisions regarding food. This reflects information from scholars who state promoting healthier eating habits is a method that could be used to prevent obesity and overweight in children (Motycka and Inge, 2007).

Since exercise and activity are incredibly important to fighting the increasing rate of obesity, I will include activities and suggestions that they can do rather than the sedentary activities most of today's youth participate in on a regular basis. Motycka and Inge also state that promoting physical activity at school can assist in preventing this epidemic from expanding. They conclude that it is important that children change the way they behave every day. Such as walking to school, taking the stairs instead of the elevator, and even how often they walk around every day as opposed to sitting down to study or watch television, all of which have a great impact on energy expenditure.

Children have the ability to make proper choices regarding their eating habits. What I find they lack are the tools to aid them in making these proper decisions. The knowledge and strength they will gain from my instructional workshops will greatly increase their awareness and ability to make conscientious decisions about nutrition and ultimately their overall health.

How do you know this will help

I am confident this course of action will help because it has been successfully accomplished elsewhere. The following information proves my theory that education is crucial to impacting the lives of children with respect to making the right decisions regarding food and nutrition as well as leading an active lifestyle.

The Nutrition and Physical Activity Program to Prevent Obesity and Other Chronic Diseases (NPAO) is based on a cooperative agreement between the Centers for Disease Control and Prevention's Division of Nutrition, Physical Activity and Obesity (DNPAO) and 28 state health departments. The program was established in fiscal year 1999 to prevent and control obesity and other chronic diseases by supporting states in developing and implementing nutrition and physical activity interventions. States funded by NPAO work to prevent and control obesity and other chronic diseases through these strategies: balancing caloric intake and expenditure, increasing physical activity, increasing consumption of fruits and vegetables, decreasing TV-viewing time, and increasing breastfeeding (CDC 2007).

They state it is vital to teach skills needed to make individual behavior changes related to nutrition, physical activity, and healthful weight – and provide opportunities to practice these skills (CDC 2007). Key words being teach skills. My plan supports this theory, and reflects the opinions of the Center for Disease Control and Prevention.

Correspondingly, it is important to create supportive environments, making healthful lifestyle options more accessible, affordable, and safe. The program provides resources, training, and mini-grant funding for schools to make changes to the school environment, including more healthful food choices. It gives them the ability to establish programs in communities to increase physical activity and/or reduce caloric intake through healthful eating habits (CDC 2007).

All of the information above proves just how vital education is to impacting ones choices. Educating the children of Roselle Park will help them make better choices in the present which will have long-lasting benefits for their futures.

Benefits

There are an incredible number of benefits that will come from these workshops. I am certain that by increasing the educational background of the children their eating habits will directly be influenced. Not only will they make better choices, but those choices will lower their potential health issues in the future. Correspondingly, the rate of all of the aforementioned health risks associating with obesity will be decreased.

Here are a few alarming facts: Children treated for obesity are roughly three times more expensive for the healthcare system than children of normal weight. The current indirect and direct costs of treating obesity have been estimated at $117 billion per year (Alliance for a Healthier Generation, 2007). One of the benefits of my plan is that the education I instill in the children will enable them to make choices to change their destiny. Ultimately, healthcare will be directly impacted in relation to the rate of obesity in adolescents as a whole.

Increasing awareness is incredibly important and will be accomplished via my proposal. Educating the adolescents of Roselle Park will help build a strong foundation for positive choices. Also, studies show that there is a correlation between nutrition and academic performance. Studies were conducted and results showed that children with iron deficiencies sufficient to cause anemia are at a disadvantage academically, unless they receive iron therapy. Improper nutrient intake can lead to anemia and other deficiencies. The study also showed that food insufficiency is a serious problem affecting children's ability to learn; and that offering a healthy breakfast is an effective measure to improve academic performance and cognitive functioning among undernourished populations (Taras, 2005). This proves that educating children on nutrition will increase academic performance.

What it ultimately comes down to is the direct impact this information and knowledge will have on the lives of the children it is being taught to. The fact that knowledge is power should be something that is confirmed and supported after review of this proposal.

Summing it up

Overall it is easy to see the many aspects as to why my proposal would be beneficial. Children all over the nation are plagued with obesity as a national epidemic. On the road to change, one must start somewhere; why not begin in Roselle Park? At 5:35pm on April 12, 2009 I will be giving an oral presentation concerning the issue of adolescent obesity in room 216 of Hickman Hall at Rutgers University. I would like to invite you to be in attendance. If you have any questions or concerns feel free to contact me. My telephone number is (908) 538-6678 and my email address is jham@eden.rutgers.edu. Email is the best way to get in touch with me because of my busy schedule, but I am available to speak on the phone any time before 5:30pm on Mondays and any time between 2:00pm and 5:30pm on Wednesdays. Do not hesitate to call or email me if you have any questions, comments, or concerns.

Sincerely,

Jane Hamilton

References

Alliance for a healthier generation. Retrieved February 7, 2009, from the website: http://www.healthier generation.org/default.aspx

Beyond boundaries. Retrieved February 5, 2009 from the website: http://www.tufts.edu/development/ news/2007/shapeup.html

Boyle, M.A. & Long, S. (2007). *Personal nutrition, sixth edition.* California: Thomson Wadsworth.

Centers for disease control and prevention. Retrieved February 5, 2009, from the website: http://www.cdc.gov/ nccdphp/dnpa/obesity/

McGuire, M & Beerman, K.A. (2007). *Nutritional sciences: from fundamentals to food.* California: Thomson Wadsworth.

Motycka, Carol and Inge, L.D. (2007). The growing problem of childhood obesity. *Drug topics,* 151(17), 33–42.

Taras, Howard (August 2005). Nutrition and student performance at school. *The journal of school health,* 75(6), 199–213.

The American physiological society. Retrieved February 1, 2009, from the website: http://www.theaps .org/press/conference/eb03/9.htm

The finance project. Retrieved February 1, 2009, from the website: www.financeproject.org/Publications/ obesityprevention.pdf

The obesity society. Retrieved February 7, 2009, from the website: http://www.obesity.org/

United States department of health and human services. Retrieved February 15, 2009, from the website: http://www.surgeongeneral.gov/topics/obesity/calltoaction/fact_adolescents.htm

Wardlaw, G.M., Hampl, J.S. & DiSilvestro, R.A. (2004). *Perspectives in nutrition.* New York: McGraw-Hill.

Jill Wu
651 CPO Way
Rutgers University
New Brunswick, NJ 08901

March 22, 2009

Rochelle D. Williams-Evans, RN, MS
Director of Health & Human Services
East Orange Health Department
143 New Street, East Orange, NJ 07017-4918

Re: Preventing childhood lead poisoning through an education program in East Orange, New Jersey

Dear Ms. Rochelle D. Williams-Evans,

As a primary member of the Health and Human Services Department of East Orange, NJ, it is noted that you and your fellow employees' work towards maintaining the safety, health and well-being of men, women, and children. The county of East Orange faces an enormous health problem: childhood lead exposure. Studies have proven that if lead poisoning is not detected early in children, high blood lead levels can cause numerous health problems such as slowed growth, hearing problems, and brain damage (*Lead in paint, dust, and soil,* 2007). These health issues, along with many more, can devastate a child's education and potential to live a productive and vigorous life. Studies have shown that out of all the municipalities in New Jersey, East Orange City contains the greatest percentage of children tested who had blood lead levels greater than the states' accepted blood lead "level of concern" (Chen, 2006). This serious issue needs to be addressed and properly solved before childhood lead poisoning reaches a level that is detrimental and irreversible.

Education and awareness of lead poisoning can be improved by using public advertisements and brochures that contain information on the dangers of childhood lead exposure, the need to get children tested, and how to prevent lead poisoning. The proper knowledge of pregnant women and parents of young children on childhood lead exposure can help increase the number of lead blood tests performed, prevent raising a child in a lead infested environment, and ultimately lower blood lead levels in children. Due to successful lead prevention programs that targeted parent education performed in the past, I feel as though my plan will help the Health & Human Services Department reach their goal of sustaining the well-being of individuals within the town of East Orange. This education program that I will introduce to you will not only save hundreds of families within this city, but also strengthen your department and the overall service to your community.

Health problems associated with childhood lead paint exposure:

There are various health and learning problems that are likely to occur when a child has lead poisoning. In the city of East Orange, children are being exposed to the dangers of lead paint. After talking about two young children in East Orange that have severe health problems as a result of lead paint exposure, Russel Ben-Ali and Judy Peet (2005) state, "The Washington children are among thousands of youngsters throughout New Jersey caught in a toxic trap" (pp. 5–6). Although the statewide blood lead "level of concern," implies that any child under 10 micrograms per deciliter of lead in blood is not at risk of any dangerous health problems, "The Centers for Disease and Control make it clear that there is no known safe level of lead toxicity" (Chen, 2006). Even after it has been proven by scientists that any amount of lead paint in blood can pose an immediate threat to the life of a growing child, East Orange still has "150 children with lead levels between 15 and 20 micrograms" (Ben-Ali, 2005, pp. 5–6).

The consequences of lead paint exposure can ultimately impair the life of a growing child. "It is well established that children under age 6 are especially vulnerable" (Jacobs & Nevin, 2006, p. 2). A special report from the Association for Children of New Jersey states that this is due to that fact that children's bodies, under the age of six, "Absorb lead more easily and at a greater rate than adults. If untreated, lead poisoning can cause behavioral difficulties, learning disabilities, retard development, seizures, coma, severe brain and kidney damage and even death" (*Eliminate Childhood Lead Poisoning*, 2007). Studies show that, "Lead paint poses health problems at any age but its potential to alter cell structure and chemistry of developing brains can devastate young children" (Ben-Ali, 2005, p. 6). In New Jersey, thousands of children suffer from the effects of lead paint exposure. There is an enormous list of permanent health effects from lead poisoning, such as mental retardation, reduced IQ, learning and reading disabilities, hyperactivity, and developmental problems in most bodily organs, particularly the central nervous system, red blood cells and the kidneys (Chen, 2006). Children six years old and under, and fetuses, will suffer the most from lead exposure because, "They are particularly vulnerable because at that age, their brain and central nervous system are still forming" (*Health effects on children*, 2004, 1). Although the symptoms of lead poisoning in children are sometimes difficult to detect, a brochure released from the United States Environmental Protection Agency states that, "The only sure way to know if a child has too much lead in his or her body is with a simple blood test. Children with high levels of lead may complain of headaches or stomachaches, or may become irritable and tired" (2004). These health problems obtained early in a child's life can seriously alter his or her education, daily lifestyle and outcome in society.

How children are exposed to lead paint:

Every year in East Orange City, men and women come together to raise children of their own, in hopes that their family will be healthy and safe. The only problem is many of these families are living in housing units that contain dangerous amounts of lead that are hazardous to the health of themselves and their children. Across the state of New Jersey, especially in East Orange, there are hundreds of occupied buildings that still contain lead. The federal government did not pass a law that banned lead-based paint from housing units until 1978; leaving hundreds and thousands of buildings, still today, contaminated with lead (*Lead in paint, dust, and soil*, 2007). Since most homes built before 1978 used lead-based paint, many families within the state of New Jersey are living in older buildings that may contain this toxic substance. Serious hazards include paint chips or paint dust on windows, windowsills, doors, doorframes, stairs, banisters, railings, fences and porches (*Lead in paint, dust, and soil*, 2007). Since children do not know any better than to put something in their mouths and chew on it, they are a lot more likely to ingest these poisonous materials and suffer the health problems.

Children and lead blood tests:

A lead blood test is the only way to insure that a child has or has not been exposed to lead. The Lead Poisoning Abatement and Control Act requires, "Health care providers to screen all children for elevated blood lead levels at both 12 and 24 months of age. Older children, up to age six, are also to be tested if they have not previously been tested, or are assessed to be high risk" (*Eliminate childhood lead poisoning*, 2007, p. 1). It is essential that children undergo a lead blood test, or a "lead screening", especially if they are living in a building that was built before 1978. Although the law does not require testing children over the age of six, tests should still be run upon previously elevated test results or other risk factors (*Eliminate childhood lead poisoning*, 2007, p. 1). Hundreds of undiagnosed children suffer each year because many parents do not take their sons and/or daughters to get tested for lead exposure. "In FY 2002, only 40% of the estimated 222,800 children between 6 and 29 months of age received a lead screen" (*Eliminate childhood lead poisoning*, 2007, p. 1). It is crucial that parents understand the importance of getting their child's blood tested for lead content. A parent's lack of knowledge of the dangers of childhood lead poisoning can be extremely detrimental to the life of a young girl or boy. According to the Centers for Disease Control and Prevention, "Increased community-wide awareness can generate broad commitment to improve community resources and political will for primary prevention" (Brown, 2005).

What educational and awareness programs like this worked in the past?

There have been several approaches across the country to try and put an end to childhood lead poisoning. In Camden City and Irvington, the Division of Medical Assistance and Health Services ("DMAHS") and the American Civil Liberties Union Foundation ("ACLU") came together to propose a project to "increase the lead screening of Medicaid-enrolled children under the age of six" (*Eliminate childhood lead poisoning*, 2007, p. 2). The groups' approach was to have trusted health care providers and agencies, such as day care employees, inform parents of the hazards of lead poisoning and the need to get children tested. Sources of education included discussions, videos, and informational packets. Results from the city agency, that overseas day care centers, showed that after more parents were informed, documented lead screenings increased by 20 to 30 percent (*Eliminate childhood lead poisoning*, 2007, pp. 2–3).

At the College of Nursing in The Ohio State University, Dr. Barbara Polivka (2006) and a team of nurses conducted a survey to determine the more favorable method of receiving lead poisoning prevention information. Results showed that, "Brochures and discussions with health care providers were the preferred methods of obtaining lead-poisoning prevention information." Younger parents, especially, preferred to acquire knowledge through billboards, brochures or speaking with someone at the health department. (Polivka, 2006). This study shows that using public advertisements would be the most favorable and influential approach to informing parents about the hazards of lead paint exposure.

Another successful project to prevent childhood lead poisoning was conducted by the Massachusetts Childhood Lead Poisoning Prevention Program (MA CLPPP). In order to educate pregnant women about the dangers of childhood lead exposure and "encourage them to adopt preventative behaviors," the MA CLPPP's project involved the, "Development and distribution of bilingual prenatal lead awareness kits packaged in large attractive diaper bags . . . including educational fact sheets and brochures, promotional items, a community resource card, an evaluation card, and a voucher for free lead-safety training for a family member" (Brown, 2005). Targeting pregnant women and expected parents can help prevent a child from ever becoming exposed to lead and suffering the consequences.

The Environmental Protection Agency (EPA) released brochures in both 2003 and 2004 containing several different topics involving how to protect you and your family from the hazards of lead. A person that is unaware of the dangers of childhood lead exposure can easily be informed after reading one of these information packets. Various topics include, "Where does lead poisoning come from?" or "Talk to your child's doctor about having your children tested for lead poisoning" (USEPA, 2004). The EPA also informs parents how exactly lead gets into the body and why "lead is even more dangerous to children under the age of six" (USEPA, 2003). These brochures provide helpful facts in a way that is informative and easy to interpret.

These models are excellent examples of how education and awareness programs can be used to reduce and ultimately prevent childhood lead exposure. If the proper education and awareness is given to men and women before they have children, more and more children can be saved from ever developing the health problems associated with lead poisoning. Preventative action needs to be taken in order to help the lives of children in East Orange, NJ.

The plan of action:

After extensive research of various techniques used to fight and end childhood lead exposure, I have formulated a plan that is bound to be efficient in solving this severe health issue. Many of the successful models above used educational programs to inform parents of the dangers of childhood lead poisoning. My program will involve the use of brochures and advertisements, containing information about the dangers associated with childhood lead exposure and the need to get children tested for lead poisoning. Centers for Disease Control and Prevention state that lead education and awareness programs should include "information about lead poisoning, evaluation and control of lead hazards, home preparation, local lead safety resources and community groups, and screening recommendations" (Brown, 2005). I will be targeting pregnant women, expected parents, and parents with young children.

I will reach out and educate men and women that are unaware of this extreme health issue occurring in their own neighborhood. Steven Marcus of the University of Medicine and Dentistry, a nationally recognized expert on lead poison treatment, states, "That would be the real gold standard- true prevention, not waiting until any kid has elevated lead level" (Ben-Ali, 2005, p. 6). Increasing awareness and improving parent education on this enormous issue can prevent a child from ever becoming exposed to lead.

Since parents of young children and pregnant women frequently visit their local pediatrician or OB/GYN, the brochures I have designed will be placed at these two types of health care facilities throughout the city of East Orange for adults and/or parents to read. In summary, the brochures will contain information on childhood lead poisoning and its various health problems, how children are becoming exposed to lead, what to do and not to do when your child is poisoned and most importantly, the absolute need to your child/children tested.

In addition to brochures planted in various health offices, I would like to reach out and inform people of East Orange through various forms of public advertisements, such as newspaper ads and billboards throughout the city. Since "many young children are not routinely screened as part of their regular physicals or 'well child' visits . . . large numbers of lead burdened children are undiagnosed and untreated" (*Eliminate childhood lead poisoning,* 2007, p. 1). As a result of this knowledge, I will be stressing the need to get children screened for lead exposure in my public advertisements.

Conclusion:

I hope that I have shown you the severity of this prevailing health issue and a possible solution to end childhood lead exposure in East Orange, NJ. I believe that you and your department agree that lead poisoning in children is a serious problem that needs to be addressed immediately. I believe my plan will not only help the lives of families within your community, but also heighten their education and awareness of the health hazards associated with lead poisoning. I would now like to take the time to invite you to my oral presentation where I will expand on my project, exploring more angles and ideas. Within my presentation, I will present my various means of advertisements and discuss the budget for my proposal. My oral presentation will take place on Monday, April 6, 2009 at 5:35 p.m., in Room 216 of Hickman Hall, Rutgers University, New Brunswick, NJ. If you have any questions or comments, please feel free to contact me at 732-463-5572 or email me at jwu@eden.rutgers.edu. Thank you so much for your time and consideration and I look forward to hearing from you.

Sincerely,

Jill Wu

References

Ben-Ali, Russell & Peet, Judy. Jersey children caught in a toxic trap. (2005, December 2). *The Star Ledger*, p. 1, 2.

Brown, Mary Jean. (2005). Building blocks for primary prevention: Protecting children from lead-based paint hazards. *Center for disease control and prevention.* Retrieved February 21, 2009, from http://purl.access.gpo.gov/GPO/LP576979.

Chen, Robert K. *Brief of Amicus Curiae public advocate of New Jersey.* (2006, April 20). Retrieved February 16, 2009, from http://www.state.nj.us/publicadvocate/reports/pdfs/Lead _Paint_Amicus_Brief_Final _2_20_06.pdf.

Eliminate Childhood Lead Poisoning: Special Report. (2004, February). Retrieved February 16, 2009, from http://www.kidlaw.org.main.asp?uri=1003&di=320&dt=4x=1.

Health effects on children. (2004, March). Retrieved February 16, 2009, from http://www.nsc.org/issues/ lead/healtheffects.htm

Jacobs, D.E. & Nevin, R. (2006). Validation of a 20-year forecast of US children lead poisoning: Updated prospects for 2010. *Environmental research,* 102, 352–364. Retrieved February 2, 2009, from Environmental Sciences & Pollution Management Index database.

Lead in Paint, Dust, and Soil. (2007, August 2). Retrieved February 16, 2009, from http://www.epa.gov/lead/ pubs/leadinfo.htm#health.

Polivka, B.J. (2006). Needs assessment and intervention strategies to reduce lead-poisoning risk among low-income Ohio toddlers. *Public health nursing,* 23, 52-58. Retrieved February 2, 2009, from Environmental Sciences & Pollution Management Index database.

U Compare Health Care. (2007). Retrieved February 2, 2009, from http://www.ucomparehealthcare.com/ drs/new_jersey/EAST_ORANGE2.html

United States Environmental Protection Agency: Office of Pollution Prevention and Toxics. (2003). *Protect your family from lead in your home* [electronic version]. *IRIS.*

United States Environmental Protection Agency: Office of Pollution Prevention and Toxics. (2004). *Give your child the chance of a lifetime: Keep your child lead safe* [electronic version]. IRIS.

Chapter 5 ■ Midterm Sales Letter Workshop

Please fill out the following form for your partner. Feel free to write comments on the draft as well.

Does the document . . .

1. directly address the funding source?   _____yes _____no

2. catch the attention of the reader?   _____yes _____no

3. discuss why the reader is appropriate?   _____yes _____no

4. include specific and descriptive headings to help guide the reader?   _____yes _____no

5. express a clear command of population and problem?   _____yes _____no

6. adequately *document* a problem for a specific location and population?   _____yes _____no

7. appropriately *quantify* the problem?   _____yes _____no

8. argue in a way that would *appeal* to the audience?   _____yes _____no

9. refer to specific *evidence*?   _____yes _____no

10. offer *examples* and/or *details* from sources?   _____yes _____no

11. cite each source in-text according to APA format?   _____yes _____no

12. describe a particular *paradigm*?   _____yes _____no

13. offer a researched *rationale* for the plan?   _____yes _____no

14. present a plan which follows logically from the *research*?   _____yes _____no

15. include the *suggestion* of a budget?   _____yes _____no

16. invite the reader to his/her presentation (including date, time, and location)?   _____yes _____no

17. include a list of *References* prepared according to APA standards with a minimum of eight published sources and at least 50% cited as print sources?   _____yes _____no

18. appear in full block form and include all six elements (return address, date, recipient's address, salutation, body, closing)?   _____yes _____no

Is the document . . .

1. signed?   _____yes _____no

2. free of all grammatical and typographical errors?   _____yes _____no

3. four to five pages in length, not including the References page(s),
   in 12 point Times New Roman font with one-inch margins?   _____yes _____no

What parts of the draft do you like the most?

What parts of the draft need the most improvement?

Additional Comments/Suggestions:

The Oral
Presentation

Chapter 6

The Assignment

The oral presentation is a ten- to fifteen-minute spoken proposal addressed to your patron (i.e., the person or people who might fund your idea). The ten- to fifteen-minute parameter does not include time spent setting up and breaking down the materials. This limit also does not include the time required for questions from the audience. This is a formal presentation, and you must use visual aids to help convey information clearly and effectively. The point of the presentation is to make a leadership statement for a specific audience that puts information into action by proposing a research-justified solution to a well-defined problem.

The oral presentation is both a useful step in the process of developing your project and a unique assignment for which you will receive a grade. It therefore serves two sometimes competing purposes:

- As an "oral draft" of the final project, it's an opportunity to rehearse your audience-awareness, to organize your research, to develop your plan, and to get feedback from the class and the instructor on how to improve your project. At least half of your grade will be based on how well you have researched your project and how well prepared you are to put together the final proposal.

- As an exercise in public speaking, it's a chance to practice the arts of oral persuasion. Part of your grade will be based on how well you perform as a speaker.

While instructors will generally focus their grades and their remarks on the strength of your content, offering advice on revision, they will also take notice of your form and poise. Usually, those students who have the strongest content do best overall.

The basic parts of the presentation are laid out in the sections that follow. I suggest that you read over the advice offered here, especially if this is the first time you have ever spoken before a group.

The Basic Parts of the Presentation

Every presentation will have to take its own form, based on the situation and the topic. If you are addressing a potentially resistant audience, for example, you may have to begin by winning them over or addressing possible objections they might have to your idea. Therefore, you should recognize that you cannot always adhere to a single form for the talk, and the outline below may have to be adapted to your particular needs.

As part of the drafting process of your proposal, the oral presentation gives you a chance to firm up your project and work out all of the parts. You should therefore keep in mind the Six P's of the project proposal: patron, population, problem, paradigm, plan, and price. Each of the Six P's should be represented in your presentation. Your talk should suggest the basic form of the final paper and should do these nine basic things:

1. Announce your topic with a "title slide," which should display your name and the title of your talk. This corresponds to your title page.

2. Begin by addressing your specific audience, explaining why they should be interested in your project. This corresponds to the letter of transmittal in the final paper, where you address the Patron.

3. Give your audience some sense of how you'll proceed, perhaps with an outline, or presentation agenda. This corresponds to the table of contents in the final paper. This could be presented on a slide, in a handout, or both.

4. Define the problem and try to quantify it in some way. This corresponds to your introduction section of the final paper, where you will generally lay out the Problem.

5. Present your research, being sure to cite sources in the proper format. This will correspond to the research or literature review section of the final paper, which is where you develop your Paradigm.

6. Describe your plan of action. This corresponds to your plan or procedures section, where you set forth the Plan.

7. Tell us about your budget and explain the Price.

8. Close with a call to action, which might correspond to your discussion section of the final paper.

9. Along the way, be sure to use visual graphic aids, just as you will in your final paper

The two main differences, then, between the oral presentation and the final paper is (1) that the oral is spoken and (2) that it is missing a bibliography of sources.

How to Prepare

As with all assignments, you'll have to prepare in the ways that have worked for you in the past. But here is some advice if you don't know where to start:

- **Research your imagined audience.** Who do you imagine might come to your talk? What is their degree of prestige and power? What level of knowledge or technical sophistication do they possess? What are their names? Many people like to begin their talk by welcoming the people in the imagined audience and thanking some of them by name for coming. This could appear on the title slide, as well. The more specifically you can imagine your audience the better your talk will be.

- **Plan ahead.** You can't wait until the last minute to prepare for a talk, and the sooner you start the better. The most important things to work on ahead of time are your visual aids, especially any visual graphic aids you want to use, such as PowerPoint slides, video and/or audio. The sooner you begin putting your materials in order, the more secure you will feel about the presentation itself.

- **Focus your talk around key points or examples.** Remember that you can't cover everything in your talk, but you will be able to cover the major points of your argument and the chief examples that support you (which you should be able to discuss in detail). If you can establish these points on paper, you will be able to focus your work.

- **Prepare an outline.** You will definitely want to prepare an outline for yourself, and you likely will want to provide your audience with an outline as well so they can follow you more easily. As you outline, pay attention to the logic and flow of your talk.

- **Develop solid visual graphic aids.** Remember one rule of thumb: if it can be represented visually, then it should be. You should have at least three visual graphic aids (visual representations of numerical information), but if your talk will cover technical information or you will be referencing numerical information you may need to use more than that. These should be effective and useful to your talk.

- **You might prepare notecards for details.** You shouldn't read your talk, but you may need to write some things down for reference. You may want to use notecards to remember numbers, names, and key details you want to cover. Number your notecards so you can keep them in order, and try to key them to your outline for easy use.

- **Know your information and examples so you can talk about them freely.** One of the best ways to prepare for the talk is just to read over your research so that you know your topic well. If you can talk about your key examples off the cuff, then you will do fine. This skill will prove to be vital in the question-and-answer part of the presentation.

- **Rehearse the talk out loud.** The key to preparing any fine performance is a dress rehearsal. Practice in front of the mirror or, better, in front of a friend. Time your talk to make sure it will not run over 15 minutes (you'll be surprised how easy that is to do), and so you have a better sense of time management. If you are especially nervous about speaking in a classroom, rehearse your talk in an actual classroom.

- **Get some sleep the night before.** A good night's sleep may be the best preparation for any situation where you will be the center of attention.

- **Double-check everything.** Make a checklist for yourself. Are your slides in order? Do you have your notecards? Make sure you have everything covered. Arrive early to test and set up any equipment you plan to use.

- **Back up all software**. You can't afford delays due to fumbling with technology. Most likely, you won't get an opportunity to reschedule your presentation.

The Question of Delivery

Delivery is all about ethos. Do we believe you? Do you impress us? Do you know what you are talking about? Like the way you package and present your final paper, the way you present your information will go a long way toward keeping their interest and attention. Here is some general advice on delivery:

- **Dress the part.** Students always ask, "Do we have to dress up for our presentation?" I usually respond, "It depends on your imagined audience." If you research your patron properly, you will know what they expect. You should definitely wear clothes that are appropriate to the context. If you want to make a good impression, it's generally a good idea to break out some of your better clothes. Sweatpants will not reflect well on you in any situation. For men, a tie is always best, but an outfit you would wear on a casual Friday at an office job might do. For women, any outfit you would wear to an office job should be sufficient. Ask your instructor for specific guidelines.

- **Create the context.** Clothes are only part of setting the stage for your talk. You'll also want to indicate your imagined audience and acknowledge their interests whenever possible. Highlight the fact that you know your imagined audience well and make sure that you keep them in mind throughout.

- **Use a tone appropriate to your imagined audience.** One way of keeping the audience in mind is by using the same language and tone that you'd use if they really were in the room. If you are asking university officials for money, for example, you wouldn't want to talk about "the RU Screw."

- **Enunciate and speak clearly.** This doesn't always mean speaking loudly, but you should speak clearly enough so that everyone can hear you.

- **Make eye contact.** Try to make eye contact with everyone in the room at some point during your talk.

- **Don't rely too much on notes.** Organize your presentation around an outline and use notecards, but *do not write out or read the presentation.* In other words: speak it, don't read it. You should know your information well enough at this point to be able to speak with confidence and knowledge using only an outline and visual aids to support and guide you. If you need to write down facts, figures, names, or an outline, use notecards because they are relatively unobtrusive. Try not to put too much between yourself and the audience...and NEVER read the slides to your audience.

- **Project energy and "sell" your idea.** If I have one major criticism of student presentations, it's that they rarely give off much energy. Imagine that you are really asking someone for money. You have to sell them on your plan. Turn any nervousness you have about the talk into energy and put a little bit of performance into your presentation. It will count for a lot with your audience and will keep them interested.

- **Ignore distractions and mistakes.** Everyone slips up here and there. Don't draw attention to mistakes, but move on so that both you and your audience can leave them behind.

- **Move for emphasis only—don't pace.** Everyone has tics and idiosyncratic actions that come out when they speak before a group. One person I know always holds a cup of water between himself and the audience as a sort of shield. Odd tics are usually an unconscious way of defending yourself from the people you're addressing. Pacing, for example, presents your viewers with a moving target so they can't hit you if they start to throw vegetables or bricks. Try to recognize these actions ahead of time and work through them. You have nothing to fear from your listeners, so try generally to stand still. Just don't stand in front of the screen too often or you'll be blocking people's view of your visual aids.

- **Be careful with humor.** Many guides to giving oral presentations will tell you to begin with a joke to loosen up your audience. What if you're talking about an especially serious topic? Use humor in moderation and only where appropriate.

Advice on Using PowerPoint Slides

Since most students rely almost exclusively on PowerPoint for their visual aids, here is some advice on preparing and using them:

- **Begin with a title slide.** Be sure to have a title slide that sets the stage for your talk and introduces yourself and your topic. It also helps to make a good first impression—especially if it is well prepared. The title slide, like a title page, should display your title, your name, and your organization. Welcome your patron and make him/her/them comfortable. Use white space, graphics, color, or design elements that are consistent with your other slides to make it attractive.

- **Use a slide for each section of your talk.** Each section of your talk—or even every topic you cover—should have its own slide. This way you can mark the turns in your argument by changing the visual image, and you can help guide your audience through each part.

- **Have one theme per slide.** Remember not to crowd too much information onto each slide. It's best to just try to cover one theme on each one. Be wary of **text-heavy** slides.

- **Give each one a header (and number them if it helps you).** Each of your slides should have its own head line or header, indicating the topic it covers. You might want these headers to correspond to the outline you presented earlier to make your talk easier to follow. Headers should have a consistent style and form and should give a good idea of what you'll cover in that section of your talk.

- **Be sure to cite sources on charts, graphs, paraphrases, or quotes.** Each visual graphic aid that uses information derived from a source should have a "source" reference at the bottom, fully visible to your audience.

- **Use large letters and a clear font.** Remember that your slides have to be seen in the back of the room as well as the front. Make them as clear and as large as possible, yet strive for an attractive appearance.

- **Maintain a consistent font and style**. All of your slides should have the same font and if you use a border it should be the same on each one. Often it is less important to follow any rule than it is to be consistent in the styles you choose. Such consistency helps to project a sense of unity to your presentation.

- **Try a unifying border or logo.** To help further project that image of unity, you can use a logo or border on each slide. This is especially useful when you are representing a company, where you may want to have your company logo or a border with colors or a style consistent with your company image.

- **Jazz it up with color if you can.** There is no question that people are impressed by color, and your presentation will stand out more if you use color in your slides and in your visual aids. However, if expense is an issue you may want to stick to black and white.

- **Strive for active voice**. Use active voice forms in your slides whenever possible, just as in all business writing.

- **Put numbers in a visual graphic form**. Remember that if something can be illustrated it probably should be illustrated. A picture is not always worth a thousand words, but it will usually keep you from using a thousand words to say the same thing. If a number or an idea or a definition or a procedure can be illustrated, it probably should be.

- **Let the audience absorb each slide.** Too often students don't leave their slides up long enough, often because they are hurrying through the presentation. Try to manage your time well and use a slide for each section of your talk, leaving each one on the screen until you raise a new topic.

- **Point to your slides for reference.** Draw your audience's attention to key aspects of your slides by interacting with them. You can do this in several ways—on screen, with the mouse, with a shadow, or with a light pen.

Some PowerPoint Slide Don'ts

- **Don't use all caps.** Studies show that people can distinguish words and parts of sentences more easily if you use both lowercase and capital letters. Readers also perceive text written in all capital letters as shouting.

- **Don't put too much information on each slide, or use long sentences, because viewers cannot absorb it all.** Try to put no more than short phrases on each slide, and don't overcrowd them. If you find yourself putting a lot of information on a slide, then likely you need to break that information up to fit on several.

- **Don't use characters smaller than 20 point.** Remember that the people in the back of the room will have trouble with small text.

- **Don't violate the rule of parallel form.** Each slide should have information that fits together in such a way that you can list it using phrases in parallel form. This helps the audience to see connections and to organize information.

- **Don't be inconsistent in capitalizing words.** In fact, don't be inconsistent about anything.

- **Don't forget to proofread for typos.** Typos on a presentation slide are like an unzipped fly: they destroy your ethos and make you look silly.

Final Words of Advice

Recognize that it's normal to be nervous.

Most people feel a bit nervous whenever they have to speak before an audience, especially the first few times they have to do so. Remember that this is normal. If fears persist, though, here are a few thoughts that might help you get past your fears:

- Remember that you know more about your topic than anyone in the room. Just try to make yourself clear and you will automatically have something to offer the audience.

- Your listeners take your nervousness for granted. In fact, since most student listeners are not used to giving presentations themselves, they expect everyone to be nervous and will either overlook or identify with your situation.

- This might be the friendliest audience you'll ever face. As fellow students, your listeners are on your side and generally want to give you high marks: I often notice that student reviewers generally see the most positive aspects of individual talks and tend to overlook problems (even after I have urged them to offer critical comments).

- Recognize that if this is your first talk it is a necessary rite of passage. The more practice you have giving presentations, the easier they will get and the less nervous you will feel each time.

- Turn fear into motivation. Nervousness can be a spur to greater preparation. Fear is not necessarily a negative thing, but the way you respond to it has to be positive. One common negative response to fear is procrastination, which is merely avoidance behavior (a variation on running away). The best response to fear is work, which can only help you in developing your project and bolstering your confidence in your subject knowledge.

If you still have worries or fears, talk them over with your professor or with friends. The more you face your fears, the better off you'll be in the long run.

Don't talk down to your audience, but challenge them to follow.

The biggest mistake that students make in presenting a technical subject is trying to get their audiences "up to speed" by giving lots of background information, usually in the form of textbook knowledge, before they begin the presentation itself. Background information should not be presented at the start, for several reasons:

- It destroys the fiction you are trying to create that you are speaking to a knowledgeable audience. Right away, you have confused your listeners as to who your audience really is.

- It sets the wrong tone, making your audience feel like they are being talked down to by a schoolmaster. Treat your audience as equals and they will prick up their ears in order to become equal to your conception of them.

- It underestimates your audience's intelligence. Because you are speaking to a college-educated audience, most of your listeners will already possess much of the basic knowledge needed to follow your talk. There may even be some audience members as expert as yourself in your field of study. Listeners will feel insulted by your explanations of "osmosis," for example, and will tune you out. Challenge them to tune in instead.

- It wastes time that you will need to present your idea. Remember that you only have a maximum of 15 minutes to give your talk. How can you present everything you learned in your core curriculum in such a short time? You can't, so don't try.

- It mistakenly tries to anticipate questions that are best left to the question-and-answer period. Remember, if someone in the audience doesn't understand something they can always ask about it afterward. And what question is easier for you to handle than the most basic questions where you get to show off the breadth of your knowledge?

- It will not make sense in the abstract. Because information is never useful except in context, audiences have a very difficult time understanding definitions, explanations, or lessons offered in report form apart from the flow of argument.

- It is unnecessary. If a presentation is organized logically, your audience will follow your argument even if they do not understand all of the details. If you feel it is necessary to explain certain technical ideas, remember that it is much more useful to offer such explanations briefly in the context of your argument (or in the question period after) than it is to give them ahead of time. Just do your thing with confidence and your audience will be impressed, especially if they don't understand all the details.

Logic should govern above all.

This final point was brought home to me once while listening to a student presentation on training co-op students to use proper care and technique in recording information in the field so as to comply with government regulations. Basically, these students were making many small mistakes (such as recording temperatures in Fahrenheit instead of Centigrade) that were destroying the integrity of whole projects. What could be more understandable? Yet the speaker began by presenting "background information" about the types of studies the students were doing and the specific data they were collecting. By the time she had finished offering that long explanation, she had to rush through her plan to train these students in better data-collecting techniques. As one reviewer in the audience noted, "I had no idea what she was talking about until she said that these students were using felt-tipped pens on rainy days to write down information." Basically, the audience did not need to know what was being written down with that felt-tip pen to understand that such pens posed a problem in the field.

State your argument up front; don't keep your audience in the dark.

You'll never have your audience's attention more than you do at the outset of your talk. So tell them as much as you can up front. Someone once said that the best advice for giving a talk is to do three things: "One: tell your audience what you're going to say; two: say it; and three: tell them what you said." While following that advice literally will lead you to an overly formulaic presentation, it does suggest the importance of leading your audience clearly through your argument with all of the forecasting statements and signposts you can muster. As I suggest above, one of the easiest ways of helping your audience to follow your talk is to provide an outline at the outset and then use slides to signal your transitions (just as you should use strong topic sentences to signal your transitions at the opening of a new paragraph in writing).

Focus on your evidence.

The most important aspect of the presentation is that you show that you have the evidence and research to support your assertions. Just as you would do in a written form, be sure to cite your sources. Name the authorities who inform your paradigm. Name the sources for all statistical data you cite. Name the authors of studies or experiments that you reference. Describe examples or models you reference in specific detail. Emphasize that there is a wide array of evidence to support you in your claims.

Illustrate your budget with a pie chart.

As part of your plan, you must include a budget, since it is one of your imagined audience's biggest concerns. Since this is one place you will always have numbers to work with, why not use a nice pie chart or other visual aid to sum up your budget? A pie chart is most appropriate because it lets you enumerate both the total and the parts.

Close with a polished call to action.

The closing of your presentation should sum up the plan you have in mind and urge your audience to act upon it. Hence the content of your close should focus on what needs to be done, and it should take a form that tries to influence your audience to act. Use whatever rhetorical powers you can muster to get them to listen. Listeners tend to remember best what comes at the beginning and at the end of a presentation more than anything in between. Therefore, in the same way you should strive to make a good first impression, you should close your talk with words that reflect well upon you as a speaker and offer up the "take home" message of your talk in a memorable way. Some speakers actually write out their closing words in order to polish and hone their form and tone. A strong close also signals clearly the end of your talk and lets the audience know it is time to applaud.

Using PowerPoint to Develop a Presentation

You are required to use visual aids in your presentation, including at least three visual graphic aids (such as graphs or charts). While there are a number of computer programs that can help you do this, no program is as effective as PowerPoint in helping you put together a coherent slide show that combines words and images.

Why Use PowerPoint?

Transparencies can be just as effective as PowerPoint slides and may be your only choice if you are not in a Smart Classroom. However, PowerPoint slides can easily be made into transparencies. In addition, PowerPoint offers both the graph-making abilities of Excel and the text-making abilities of Word while giving you powerful tools for keeping your slides consistent in layout and design.

What to Include on Your Slides

Do not try to put everything you want to say on your slides or you will overwhelm your audience with information. Instead, your slides should emphasize the major points and primary evidence that you want your audience to remember. Focus your PowerPoint slides around key points or examples. Ideally, you should try to limit each slide's content to four to five bulleted points (never more than seven) that are about five words each (the sound-byte version of your talk). This way your audience will be able to focus on what you are saying rather than focusing on reading the slides.

Getting Started

Open PowerPoint. You will automatically see this screen once the program loads:

Because writing and presenting is an active decision-making process, the best presentations will usually begin by selecting either the "Design Template" or "Blank Presentation" option. This way you can control the content-making process so that your presentation best suits your audience's needs. The design template offers you a professional-looking presentation. However, these templates are somewhat generic and are usually recognized by business professionals, so you might consider creating your own template by selecting "Blank Presentation."

For a blank presentation, *select "Blank Presentation" and click OK.* Now you can see the slides that are available to create a presentation.

Creating a Title Slide

You should begin your presentation with a title transparency that sets the stage for your talk and introduces you and your topic. A good title slide will also help you make a good first impression. The title slide, like a title page, should display your title, your name, and your organization. Use white space, graphics, color, or design elements that are consistent with your other slides to make it attractive. Any of the blank presentation slides will work as a title slide, but the first slide choice is most commonly used.

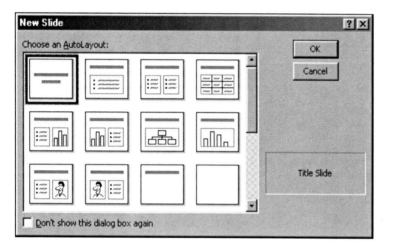

Highlight the type of slide you would like to use. Click OK. Now you can edit the slide.

When you have finished, *be sure to save your work!* Once you have saved, *start a new slide by clicking on "Insert" and then scrolling to "New Slide."*

Organizing the Rest of Your Presentation

Before you select the blank slides for the rest of your presentation, think about the logical progression of your talk. You might find the Six P's a useful basic outlining tool, moving in order through Patron (why did you choose this audience?), People (who needs help?), Problem (what evidence do you have of a problem?), Paradigm (what research informs your solution?), Plan (how will your plan be implemented?), and Price (how much will it cost?). Try to chunk major ideas together on each slide, outlining your talk so that the audience can remember its most important points and understand the parts of your argument. For each section, ask yourself what the key ideas are and develop bullet points to represent each one. And spend some time thinking about what visual graphic aids will be the most powerful in persuading your audience.

You should especially try to use visual graphic aids to help quantify the problem, since a picture can be a powerful persuader early on in your talk.

Inserting Tables

Tables are used to show a large amount of numerical data in a small space. They provide more information than a graph with less visual impact. Because of the way they organize information into vertical columns and horizontal rows, tables also permit easy comparison of figures. Be sure to use concise, descriptive table titles and column headings. In addition, arrange the rows of the table in a logical order. When putting dollar amounts into tables, be sure the decimals are aligned for easy addition or subtraction. *To create a table, click on the fourth slide.*

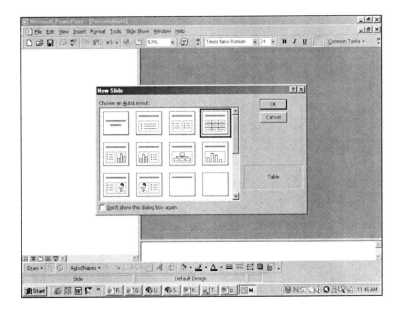

Before PowerPoint will insert a table, it asks you to enter the number of columns and rows you need. *Enter the number and click OK.*

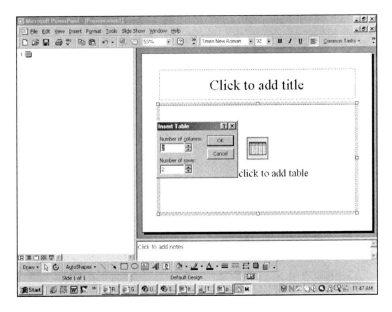

Through the Tables and Borders toolbar, PowerPoint enables you to align text with the top or bottom, center text vertically, change the border color, change the cell color, and manually split cells with the "draw table" button.

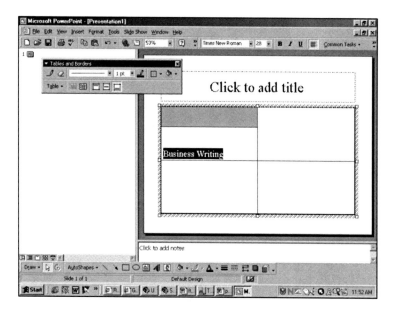

Inserting and Editing a Graph

Before you make a graph, think about the information you want to convey. Draw a picture by hand and think about whether there is room for text on the page as well. *Choose a new slide that has a graph where you want it (with text or without)*; you will see a spreadsheet and a preformulated graph. If you don't see the spreadsheet, double-click on the graph and it will appear.

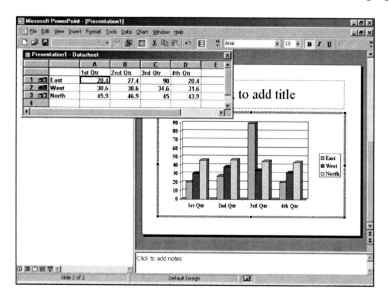

PowerPoint's graph-making layout is like that of Excel. You do not draw the graphs yourself. Rather, you change the data in your spreadsheet and then PowerPoint draws the graph you need automatically. However, you can change the appearance of your graph directly. Click on any part of the graph to change its size, shape, or color.

To change the type of chart you are using, click on "Chart" and scroll to "Chart Type." You will see the following menu where you can choose several different graphing options, including bar graphs, pie graphs, line graphs, and even bubble graphs. *Select the graph that is best suited to conveying your data.* For instance, line charts are generally used to show changes in data over a period of time and to emphasize trends. Bar charts compare the magnitude of items, either at a specific time or over a period of time. Pie charts compare parts that make up the whole. With pie charts, you should start at the 12 o'clock position with the largest unit and work around in descending order.

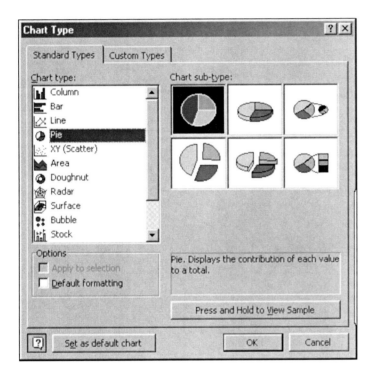

Whatever chart you choose, be sure to keep it simple. Your goal is to focus the reader's attention on the data in the chart rather than on the chart itself. In addition, you will want to label all charts as figures and assign them consecutive numbers that are separate from table numbers. This way if your audience has a question about a specific image later, they can refer to it by name and number.

Using Photos and Graphics

The "power" of PowerPoint comes from the way it allows you to seamlessly integrate text and images. Images can be a powerful support for your message, though they can also distract from what you want to say if they are not well chosen. Ideally, you will want to develop your own original images to use in your presentation. But sometimes noncopyrighted graphics and images can be useful. *To insert clip art, simply double-click on the image field and follow the screen instructions to choose items from the Microsoft Clip Art Gallery. To add an image you have saved on your computer, click on "Import Clips" and navigate to where you saved it. To include photos or graphics from a website, you can first save them on your computer or simply right-click on the object and then click on "Copy."*

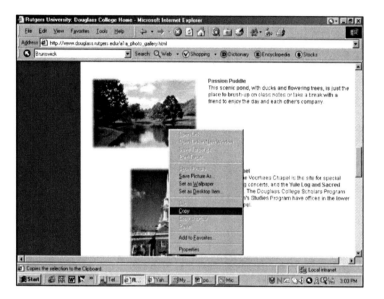

Return to your PowerPoint presentation. Right-click on the slide where you want the object and select "Paste."

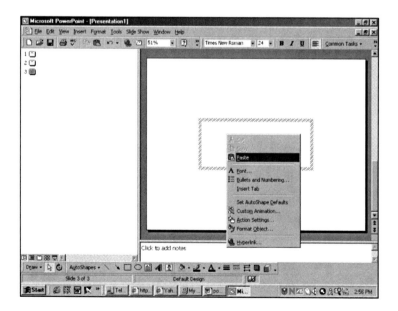

Making a Master Slide

You can also make any graphic or photo into a background for all of your slides. *To create a background go to "View," scroll to "Master," and click on "Slide Master."*

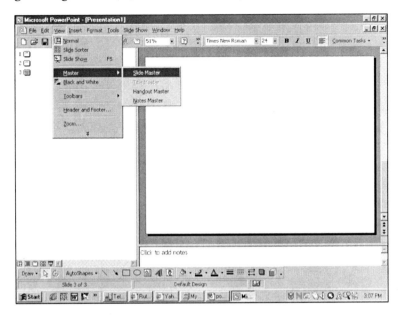

Right click on the center of the slide and select "Paste."

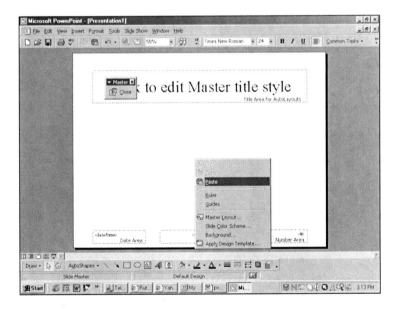

Click on the graphic. Grab a corner of the graphic and stretch the object across the screen so that it will fill the background of your master slide.

Sometimes it is hard to see text against an image background. But if the image is well chosen or especially appropriate to your topic it can help you to create an original background that breaks from the familiar generic backgrounds that most presenters use. Also, there are ways to make a background image fade into the background so that your text and support graphics can take center stage.

To blend your image into the background of a master slide, try the following:

Click on "View." Scroll to "Toolbars." Check "Picture" and "Master." Alternate between clicking on "Less Contrast" and "More Brightness" until your background is light enough so your audience can read the text on your PowerPoint slide.

Click on "Close" on the Master toolbar to apply this background to all of your slides.

Adding Text

Once you have the basic visual layout of your talk you can add your text. This part is easy. *To make a slide with bullet points select the second slide from the New Slide menu.*

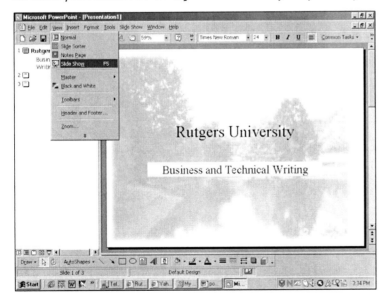

Add the text that you want.

Viewing Your Presentation

To view your presentation, go to "View" and click on "Slide Show." Press the right arrow key to advance your slides. *To exit, press the Esc key on your keyboard.*

You are now ready to give a PowerPoint presentation!

Creating Your Own PowerPoint Template

This brief tutorial will guide you through the process of creating your own template for your PowerPoint slides, which will help individualize your presentation and help it stand out from others.

STEP 1: Find, create, or modify an image using Adobe Photoshop.

Unless you have a lot of experience with Photoshop, you will most likely want to find and modify an image to serve as your background. Therefore, I will focus on that option.

1A. Use Google Image Search to find something suitable.

In my own search, I went to Google Image Search (http://images.google.com/) and tried the term "leukemia." I quickly located an interesting image here: http://www.stanford.edu/group/cleary/leukemia.jpg

WARNING: Be sure to pick an image that is not too busy and will not distract from the information you present. Try out your template image on other people before you commit to it. Also, if you are doing a presentation in a professional setting, you may want to find noncopyrighted images or create your own images to avoid problems of copyright infringement.

1B. Save the image to your My Pictures folder.

1B1. Get the best version of the image onto your computer's browser window.

1B2. Right-click on the image and choose "Save Picture As . . ."

1B3. Save the image to your My Pictures folder (usually the default setting) or anywhere you can easily find it and work with it later. The My Pictures folder, by the way, is located in your My Documents folder.

1C. Modify the image with Photoshop to create an 800 x 600 pixel .jpg or .gif file.

1C1. Open Adobe Photoshop.

Go to Start → Programs → Adobe Photoshop CS

1C2. Use Photoshop to open your image.

In Photoshop, go to File → Open → My Pictures → and navigate to your image. If you have the image open on your desktop, you can also simply drag and drop it into Photoshop directly.

1C3. Change the width or height of your image to > 800 × 600 pixels.

Go to Image → Image Size and change the width or height accordingly. In my case, I had to modify the width to 800 pixels and the height was greater than 600 automatically.

1C4. Create a New file 800 × 600 pixels.

Go to File → New and use the New window to create an 800 × 600 pixel background.

WARNING: Be sure to adjust the background and Color Mode accordingly. You will probably want to use "RGB Color."

1C5. Drag the ">800 × 600" image from its pane over to your new 800 x 600 file pane.

Be sure that the image is placed properly. You can also now modify the image further, as I will discuss. For example, to make an area where your text will display well, you can use the Eyedropper tool to pick the dominant color of your image and then the Paint Brush tool (with a large-sized brush, and the opacity set to about 50%) to dim your image slightly and make it fade into the background. This way your text will be easier to see later. There are other ways to modify your image, including the Clone Stamp and other tools. But painting an opaque layer over it is the easiest and fastest way to get a more simplified background to help your text stand out.

1C6. Save for Web.

Go to File → Save for Web. Save the image as a .jpg or .gif file to the My Pictures folder or wherever you will be able to find it easily later. This is the image you will use as your background.

WARNING: Be sure to select either .gif or .jpg in the Save for Web dialog box. Also, using Save alone will not work for you since Adobe will only save your working file as a PSD file, which you cannot use later. Be sure you are saving a .jpg or .gif file.

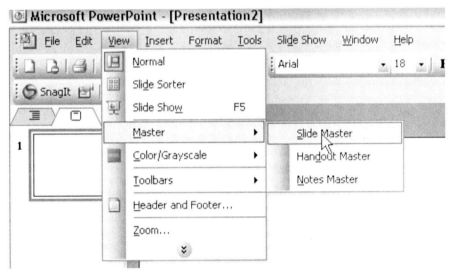

Save for web

1C7. Close Photoshop and Save your work files (.PSD) just in case.

STEP 2: Create the PowerPoint Template

2A. Open PowerPoint.

2B. Begin a new Blank presentation.

Choose File → New. The New Presentation task pane appears. Click the New → Blank presentation option on the right of the screen.

2C. Open the Slide Master.

Choose View → Master → Slide Master.

2D. Choose Insert → New Title Master.

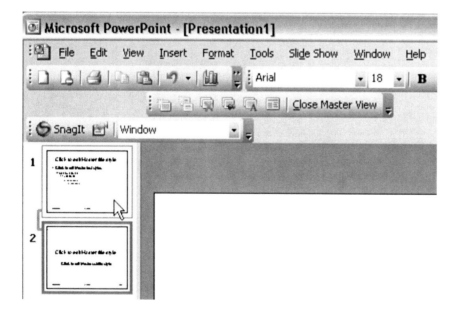

You will see two slide previews in the left pane. Both previews are linked through a connector. The top preview represents the slide master and the bottom preview represents the title master. Any elements you place within these masters show up on all of the slides based on them. The title layout slides are based on a title master. All of the other slide layouts are influenced by the slide master. (PowerPoint has more than 26 slide layouts, which you will find in the Slide Layout task pane.)

2E. Add your background image.

Choose Format → Background.

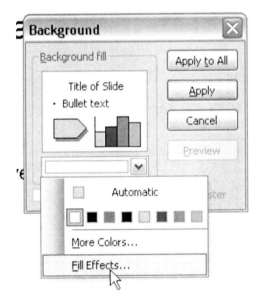

2E1. In the Background dialog box, click the downward-pointing arrow and click "Fill Effects."

2E2. Select the Picture tab.

2E3. Find your picture (probably in My Pictures) and insert it.

2F. Save Your Template

It's a good idea to save your template before we go further as a .pot (Design Template). Go to File → Save and be sure to select "Designer Template" or "PowerPoint Template" (.pot) from the "Save as type" part of the dialog as shown.

Once your design template is saved, you can create a new presentation using it.

Adapted from:

http://www.computorcompanion.com/LPMArticle.asp?ID=197

See the rest of that tutorial for additional help.

Two Examples of PowerPoint Presentations

Slide 1

Enhanced Transportation for the RU Community

The New Rutgers Car Pool Program

Slide 2

Overview: Carpooling, Why Now?

- Time is our enemy
- Who is affected?
  - Students and faculty of Rutgers University
  - All workers, and citizens of New Brunswick
- This presentation, details a proposal to enhance our transportation options
  - The problem
  - The research
  - The plan and its cost
- Now is the time to establish a carpool program at Rutgers University

Slide 3

The Problem and Who it Affects

- Congestion, pollution and not enough parking
- Inconvenience
  - To students and faculty getting to class on time
  - To business people and students getting to work with minimum hassle
- An annoyance for the entire New Brunswick and surrounding communities

Slide 4

We All Need to be concerned about Emissions and How it Relates to our Driving Habits

Slide 5

The Research

- Rutgers already has one of the largest bus transportation systems
- A carpooling program is intended to enhance, not reduce use of public transportation
- Enhancing options can be a solution

Slide 6

Are Students Ready To Car Pool

Slide 7

A Model of Success

- University of Maryland
  - Smart Park Carpool Program
  - Promotes environmental conservation and for students and faculty to be responsible citizens
  - Usage is monitored
  - Enhances bus system and promotes all forms of alternative transportation

Slide 8

The Plan: Key Actions

- Develop an internet based system
- Install card readers to monitor system
- Create promotional material to advertise and promote the system
  - Provide incentives to maximize participation
- Develop system to fund ongoing costs

Slide 9

Estimated Budget

- Develop and purchase software -$15,000
- Advertising Campaigns and Incentives- $15000
- Card Readers- $5000
- Emergency Ride Home -$5000

Initial Estimated Budget $40,000

Slide 10

Funding and Ongoing Management

- "Rutgers Riders" program membership fee
- Tiered parking fees for participants and non-participants
- Excess emergency ride home fee
- Student-Faculty over site team

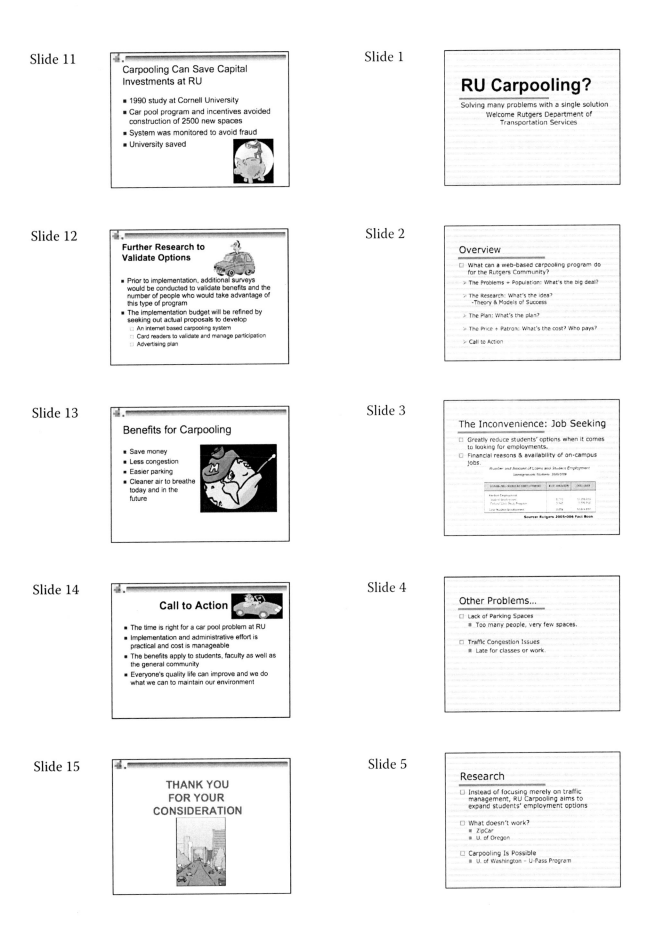

Slide 6

What doesn't work?

- ☐ **ZipCar**: The Good and Bad
  - ■ Gives students more mobility for occasional off-campus errands
  - ■ Financially unrealistic to drive on a regular basis for jobs/internships

- ☐ **U. of Oregon's Carpooling Program**
  - ■ Available only to faculty members
  - ■ Very low participation rate in program due to:
    - ☐ Not enough publicity
    - ☐ Lack of incentives

Slide 11

Estimated Budget

Website & Database Development	$5000
Advertising & Marketing Campaign	$8000
◆ Flyers, Posters	
◆ RU-TV Commercials	
◆ Imprinting on Parking Tickets	
Cost for Participants' Incentives	$15000
◆ Discounts on parking permits	
◆ Emergency Ride Home Programs	
◆ Free gas cards	

Total Cost for First Year: $28000

Slide 7

Model of Success:
University of Washington

- ☐ U-Pass Program
  - ■ Aims to encourage alternatives to driving alone to/from the U. including <u>carpooling</u>

  - ■ Award-winning and nationally recognized as a successful Transportation Management Program.
    - ☐ Governor's CommuteSmart Award
    - ☐ EPA and Department of Transportation: The Commuter Choice Leadership Initiative

Slide 12

Operating Funding

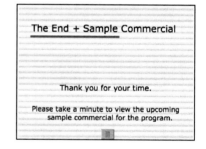

Slide 8

Statistical Evidence

- ☐ Declined percentage of drive-aloners

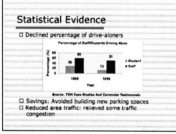

- ☐ Savings: Avoided building new parking spaces
- ☐ Reduced area traffic: relieved some traffic congestion

Slide 13

Call to Action

Carpooling makes sense!

Slide 9

The Plan: RU Carpooling?

- ☐ Website with searchable database exclusive for Rutgers students/faculty members

- ☐ Incentives & Benefits
  - ■ Receive 3 "daily passes" each month for free
  - ■ Emergency Ride Home or Reimbursement on traveling cost
  - ■ Discount on parking permits
  - ■ Reserved parking lots

Slide 14

The End + Sample Commercial

Thank you for your time.

Please take a minute to view the upcoming sample commercial for the program.

Slide 10

The Plan (Continued)

- ☐ Cost to Participate
  - ■ $20-$40 per semester depending on if the participant will be sharing the driving/providing a car.

- ☐ Marketing
  - ■ RU-TV Commercials
  - ■ Flyers/Emails
  - ■ Advertisement on Parking Tickets

The Science of PowerPoint Overload

Cliff Atkinson

The PowerPoint landscape has changed with the research of Richard E. Mayer of the University of California, Santa Barbara.

Rich's work is not about PowerPoint. It is about the way the human mind works. And with that core understanding, his research questions explore the best ways to present information that align with the mind's processing capability and capacity. Because of the way he approaches his research, his findings apply to a range of multimedia, including PowerPoint.

What does this mean for PowerPoint users? It is time for the sacred cows of PowerPoint to be sacrificed on the altar of scientific research.

For example, it is conventional wisdom to put no more than five lines of text on a slide, with no more than six words per line. But that convention is no longer wise in the light of research that shows that putting so much text on a slide is a recipe for information overload. As Rich said in a recent interview at Sociable Media:

"Cognitive scientists have discovered three important features of the human information processing system that are particularly relevant for PowerPoint users: *dual-channels*, that is, people have separate information processing channels for visual material and verbal material; *limited capacity*, that is, people can pay attention to only a few pieces of information in each channel at a time; and *active processing*, that is, people understand the presented material when they pay attention to the relevant material, organize it into a coherent mental structure, and integrate it with their prior knowledge."

In this light, a screen full of bullet points overloads the visual channel beyond its capacity, leaving little time for integration to happen.

A solution? Reduce visual overload by moving text off-screen, and shift processing to the auditory channel by narrating the content instead. A practical solution in PowerPoint is to design a "slide" in the Notes Page view, placing written explanation in the off-screen Notes section below, and using the on-screen area above for an illustration and a few descriptive words. This solution offers a better projected media experience, plus more comprehensive handouts when the PowerPoint is printed in Notes Page format. More work for you? Yes. A better learning experience for your audience? Yes. The reality is that we have to work harder to make PowerPoint easier for people to understand.

Rich's research marks a PowerPoint turning point. The science is there, and now it is up to us to deny it or accept it and change. If we decide to change, the change needs to happen in our own minds and behavior. Why do we put all those bullet points up there anyway? For some people it is to help them structure their thoughts, and to remind themselves of what they want to say. For others, it is to cover themselves by providing a record that they told somebody something: "See, it's there in the PowerPoint." But in any case, none of these reasons address the issue of the proper form the information needs to take for someone to process it effectively. And that is what we all need to learn how to do.

You can read more about Rich's research in his book *Multimedia Learning*, or in his paper with Roxana Moreno of the University of New Mexico in *Educational Psychologist* titled "**Nine Ways to Reduce Cognitive Load in Multimedia Learning**." If you're short on time, here's a "Cliff's Notes" version of some of Rich's research-based design principles as they apply to PowerPoint:

Do your slides contain only words? Show some pictures.

Do your slides contain words that you also speak? Stop being redundant.

Do your slides contain things you don't explain? Get rid of them.

Do you pause for a long time on a single slide? Break it up into smaller pieces.

Do you have lots of information on a slide? Keep it simple.

Beyond Bullet Points:

How to unlock the story buried in your PowerPoint

Cliff Atkinson

■ ■ ■

If you use bullet points in your PowerPoint presentations, it's probably because writing bullets helps you to build slides quickly and reminds you to cover all the points you want to make. But although bullet points may help you to do many things, one thing they cannot do is help you to tell a story.

Some of the world's largest organizations have adopted the word "story" as their new mantra for corporate communications. Marketing messages should tell a story, corporate strategy should tell a story, mission statements should tell a story, and even Web sites should tell a story. Why the sudden interest in stories? For one clue, look no further than the approach you may be applying to your own PowerPoint slides, which locks out the possibility of telling a story in the first place.

Bullet Points

The origin of bullet points in presentations is actually clearly visible on most PowerPoint slides—a type of outlining approach that everyone uses, yet no one questions. This approach always begins by placing a category heading at the top of a slide—such as Our History, Challenges, Outlook and Lessons Learned. It is remarkable that you see exactly the same headings in every presentation, across organizations, professions and even cultures.

These headings do nothing more than establish a category of information, which you then explain with a bulleted list below it. Although this approach can help you create slides quickly, it also guarantees that you never do anything more than present a series of lists to an audience. When the primary way we communicate is by presenting lists to one another, it is no wonder that the phenomenon of story is gaining momentum, because a story is the opposite of a list. Where a list is dry, fragmented and soulless, a story is juicy, coherent and full of life. Presented with the choice, any audience will choose life.

So, that leaves us with the essential problem: If we can agree that the era of the story is dawning, and that bullet points are standing in our way, how do we unlock the power of a story in our PowerPoint presentations? This is becoming an issue of strategic concern to major players by large organizations where PowerPoint has replaced the written word as the predominant way of communicating information.

What Kind of a Story?

The concept of a story may be new to the boardroom, but storytelling is at least as old as the person who defined it as an art 2,400 years ago: Aristotle. If you think Aristotle's ideas on story are no longer relevant, look no further than a movie screen. Hollywood screenwriters still credit Aristotle with writing the definitive elements of story, including action, a plot, central characters and visual effects.

But even Aristotle knew that not all stories are created equal. So, the natural question arises, "Exactly what kind of story is appropriate for a presentation?" For example, a story can take the shape of a Hollywood blockbuster meant to entertain, or a story can be a colorful anecdote about something that happened on vacation. Although both are stories, neither is complete enough to fulfill the complex needs of presenters and audiences today, both of which need much more than entertainment or personal anecdotes in order to make fully informed decisions.

Instead, we now need a specific type of story that blends a classic story structure along with classic ideas about persuasion. Again, Aristotle adds a great deal to the discussion, because he wrote the book on persuasion, in addition to the book on storytelling. In order to bring Aristotle's ideas up to date in a media-savvy world, you need to blend one part storytelling, one part persuasion and one part Hollywood screenwriting to create a powerful approach for your PowerPoint presentations.

Unlocking the Secret Code of a Persuasive Story

A persuasive story uses the structure of a story, but spins the story in a particular way that ensures it aims at achieving results we need in presentations: by using persuasion. You can apply this fundamental structure to any type of presentation. Using a visual medium such as PowerPoint gives you additional levels of communicative power, the same ones that Hollywood shows us every day.

For example, let's see how a persuasive story looks in the form of the first five slides in a PowerPoint presentation to a board of directors, where the presenter is seeking approval for a new product. Instead of using a category heading, the top of each slide features a simple statement that addresses each category of information that the board needs to know about, as described here.

Slide 1: Establish the setting. The headline of Slide 1 reads: Our sector of business is undergoing major change. The subject of this headline establishes the common setting for the presentation and relates the "where" and "when" for everyone in the audience.

Slide 2: Designate the audience as the main character. The headline of Slide 2 reads: Every board faces tough decisions about what to do next. The subject of this headline establishes the members of the board as the main characters in this story, establishing the "who" of the story.

Slide 3: Describe a conflict involving the audience. The headline of Slide 3 reads: Six new products have eroded our market share. The subject of this headline describes a conflict the board faces that has created an imbalance. This explains "why" the audience is there.

Slide 4: Explain the audience's desired state. The headline of Slide 4 reads: We can regain profitability by launching a new product. The board doesn't want to stay in a state of imbalance, so the subject of this headline describes the board's desired state, describing "what" the audience wants to see happen next.

Slide 5: Recommend a solution. The headline of Slide 5 reads: Approve the plan to build Product X, and we'll reach our goals. This final headline recommends a solution, describing "how" the audience will get from its current state of imbalance to its desired state of balance.

Reading these five headlines in succession reveals an interesting and engaging story that will be sure to capture the board's attention. And when you add an illustration to each of these headlines, you open up the power of projected images, including full-screen photographs, clip art, or even simple animated words.

The Rest of the Story

The five slides in this example form the backbone of Act I of a persuasive story structure. Act II then spins off of the pivotal fifth slide, explaining the various reasons why the audience should accept the solution. Act III frames the resolution, setting the stage for the audience to decide whether to accept the recommended solution.

With the solid structure of your first five slides in place, your presentations will move well beyond the stale world of bullet points, and into the lively world of a persuasive story. By blending together the classic concepts of story and persuasion with your PowerPoint software, you are sure to engage your audience and make things much more interesting—and productive—for both you and your audience.

Graphic Aids Assignment

Visual aids are an important part of both your oral presentation and your final project. They can provoke an immediate response in your audience in a way that a paragraph of statistics may not. In preparation for your oral presentation, bring in at least three visual aids with written commentary.

For the purposes of this assignment, these three visual graphic aids should be taken from popular sources, such as newspapers or magazines, or printed out from online sources (e.g., Google images). They could be ones that you are considering using for your oral presentation and project proposal, but they could also just be interesting examples of graphics. Please do not bring in pictures or photographs; I would much rather have you find images which are visual representations of statistical information, as in charts, tables, graphs, and so on. You may find certain graphics that you find misleading, or would like to show the class possible "sneaky" tactics used by the presenters. To help you develop good, informative, attractive visual aids, we will look at your examples and some others in a peer review/class presentation session.

Name: _____ **Course:** _____ **Semester:** _____

■ Oral Presentation Evaluation

1. **Audience:** How well did the speaker address the funding source?

   1   2   3   4   5   6   7   8   9   10

2. **Eye Contact:** How well did the speaker acknowledge and address those actually present?

   1   2   3   4   5   6   7   8   9   10

3. **Delivery:** How were the speaker's volume, enunciation, posture, appearance, and body language?

   1   2   3   4   5   6   7   8   9   10

4. **Evidence:** Did the speaker support claims, give examples, reference facts, and cite sources?

   1   2   3   4   5   6   7   8   9   10

5. **Organization:** Was the presentation easy to follow? Were all Six P's represented, in the correct order?

   1   2   3   4   5   6   7   8   9   10

6. **Visual Aids:** Were there sufficient, attractive, and useful visual graphic aids?

   1   2   3   4   5   6   7   8   9   10

7. **Preparation:** Did the presentation show careful planning, good time management, and smooth transitions?

   1   2   3   4   5   6   7   8   9   10

8. **Questions:** Did the speaker demonstrate knowledge, confidence, courtesy, and interest?

   1   2   3   4   5   6   7   8   9   10

Additional Comments/Suggestions

The Project Proposal

Chapter

The Assignment

The project proposal is the final draft of the project you have worked on all term. Like the oral presentation, it should be a leadership statement that puts information into action by proposing a research-justified solution to a well-defined problem. Unlike the presentation, though, it must adhere to a specific format, which is presented below and illustrated in the sample papers that follow. The guidelines for preparing this final paper may not conform to those of your workplace or those requested for specific grant applications you might be considering. These guidelines, though, should be readily adaptable to any real-world submission. We encourage you to revise your final project for submission in your workplace or in your future graduate work, but for the time being focus on fulfilling the requirements of our class. Please consult with your instructor if there are any discrepancies between the parameters presented here and the instructions included in a published Request for Proposals.

Remember that the heart of the proposal is a problem, paradigm, and plan that work together to create a unified concept. The paradigm should grow organically out of the way you define the problem, and the plan you present should be clearly rationalized by the paradigm. If you unify and focus your argument, you will be able to present a well-organized and logical paper.

The final draft of the project proposal should be from fifteen to twenty pages inclusive, single-spaced (though your References should be double-spaced in keeping with APA guidelines). You should also try to do the following:

- Strive for a consistent professional tone throughout.
- Number your pages clearly.
- Provide coherence to your paper using rhetorical, design, and signposting strategies.
- Use clearly distinguished headings and subheads to help guide your reader through the parts of each section.
- When appropriate, use bullets or numbers to list items for easy comprehension.
- Label and number all graphs and figures for easy reference.
- Unify your paper with a consistent typography and style.

- Polish your writing for style and emphasis.
- Proofread for errors in spelling, grammar, and syntax.

The Parts of the Proposal

The formal aspects of the final proposal help you to present your overall argument in a way that is useful for your reader. There are thirteen parts of the project proposal, most of which should be labeled and presented in order (with the exception of visual graphic aids, which should ideally be incorporated into the body of the paper with individual titles):

1. Cover Letter—generally one full page (not numbered or titled)

2. Title Page—one page (not numbered or titled)

3. Abstract—one page (Roman numeral i)

4. Table of Contents—one page (Roman numeral ii)

5. Table of Figures—one page (Roman numeral iii)

6. Introduction—generally more than two pages (Arabic numeral 1+)

7. Literature Review (or Research)

8. Plan (or Procedures)

9. Budget

10. Discussion (perhaps including an Evaluation Plan)

11. References

12. Visual Aids (or Figures)—incorporated into the text when possible

13. Appendix (if necessary)

1. Cover Letter

Like the cover letter that accompanied your résumé, this letter of transmittal is intended to explain and interpret the attached document. It should explain why the reader has received your proposal, and it should try to persuade the reader to examine it closely, offering details about the content intended to interest or intrigue him or her. The letter of transmittal should respond to the situation of reading and answer the reader's likely questions: "Why is this on my desk?" and "Why should I read it when I have a dozen other things to do?"

The transmittal letter can take the form of a letter (for a reader outside of your organization) or memo (for a reader within your organization). While an increasing number of transmittals are written in e-mail form, where the proposal is usually an attached file, we ask that you adhere to the traditional paper forms for the purposes of this course.

If it is a letter, it should follow the full block style, in which all of the elements are flush with the left margin in this order:

1. Return address (your name and address)

2. Date (for the purposes of the class, use the due date of the final proposal)

3. Recipient's address (including name, title, organization, and business address)

4. Salutation ("Dear" plus formal address and name)

5. Body (see discussion below)

6. Closing ("Sincerely") and signature

If you are using the letterhead of a specific organization, you will not need to include your address. If the cover letter is prepared as a memo, then it should be written on company stationery (or facsimile) and prepared in memo form:

- To: (addressee's full name)

- From: (your full name and handwritten initials)

- Date: (today's date)

- Subject: (a line indicating your proposal topic)

- Body (see discussion below)

Many of the rules for writing the cover letter to accompany your résumé obtain here. Since your imagined reader probably attended your presentation (or at least you created a context where he or she was imagined in the room), you may want to begin by reminding the reader of that event, explaining that this is the full version of that proposal. Whether or not you have met your reader before, begin by explaining why you sent him or her your proposal and why it should be of interest. Emphasize what you know about the reader's interests, and highlight the principal way in which this proposal matches those interests.

The central paragraph (or central two paragraphs) should offer an overview of the project, highlighting salient details about the problem, paradigm, and plan. Again, point to those aspects of your project most likely to interest your reader.

The final paragraph should invite further contact, offering the most convenient way for the reader to get in touch with you (perhaps by phone or e-mail).

2. Title Page

The page should include the following information:

- Project title

- Submitted by: Your full name and title (or position)

- Submitted to: Your addressee's full name, title, and business address

- Date

You should also indicate somewhere near the bottom of the page the course for which this paper was prepared, your instructor's name, and any class information requested by your instructor. (This way if your paper gets lost it won't end up on the desk of the imagined audience but will have a chance of getting returned.)

The title of your project should be carefully chosen and crafted for maximum communication in the shortest space. It is one of the first things the reader sees of your report, and it will become the means of referencing it to others. The more communicative power it has, the more effective it will be. Strive to be both clear and memorable. Remember that you can use a two-part title, especially if you want to give your project a catchy title followed by a more technically specific one.

There are many ways to design the title page, and you should do what looks and works best for your specific project. Use white space, color, and other page elements to design an attractive image that is consistent with the document design as a whole. You might want to use graphics or pictorial lettering to highlight your topic.

3. Abstract

The abstract should be clearly labeled as an "abstract" at the top of the page and should be no more than one or two paragraphs in length. The purpose of the abstract is to tell busy people (or their secretaries) how to file your report. It should be written from a disinterested perspective, providing a balanced view of the project idea as though written by an outside party. Usually it is written in the third person or uses passive voice to avoid naming the agent. For the purposes of this class, you should write a relatively long, informative abstract that includes details about your overall argument and covers elements of the problem, paradigm and plan (in that order). Be sure to indicate your rationale and what specific action you want to take. Aim to be maximally communicative within minimal space—generally between 150 and 300 words.

4. Table of Contents

Clearly label and design your table of contents for easy use. Recognize that the table of contents has two main uses: it helps readers locate the information that interests them most (this is especially true of long reports) and it gives your reader an overview of the project and its parts. You should list all parts of the project listed above (excluding the cover letter, title page, and visual aids), along with any important subheads. Number the opening parts (abstract, table of contents, table of figures, and executive summary) with small Roman numerals (i, ii, iii) and then use Arabic numbers (1, 2, 3) beginning with the introduction section. Use whatever design elements you can to help make the information clear and usable—indenting subheads, using ellipses to link section names and page numbers, and aligning all related parts. The style and font should be consistent with the design throughout your document.

You can work up a table by carefully laying out the items in it, but many word processing programs, such as Microsoft Word, will generate a table for you.

5. Table of Figures

If your table of contents is short, you might include your table of figures (clearly labeled) on the same page. Otherwise, it should occupy its own page. Ideally, each figure and illustration you use should have a number for easy reference. List the number and title of each figure along with the page on which it appears.

6. Introduction

There are two purposes for the introduction: to present information about the problem you will address and to forecast your overall argument. Here is where you will want to offer all the information you have on the problem you seek to address. You should try to quantify or define the problem and offer images that help clarify and emphasize the key aspects of it. Focus on those aspects of the problem that will most interest your reader, and suggest by the way you examine or define the problem a direction for approaching it. Close the introduction with a forecasting statement giving your reader a sense of your argument to follow and providing a transition to the next part.

7. Literature Review (or Research)

This is the section in which you present, analyze, and integrate your paradigm research into your proposal. The literature review section should open with some reference to the problem (especially by way of transition from the introduction), but it should focus mostly on the justification for your project. The research you present should explain why you will approach the problem in a particular way; it should also provide a unified rationale for the specific plan of action you describe in your plan. Thus the paradigm is essential for unifying your paper because it shows how the plan of action you will propose is a logical approach to the problem you have defined. Remember, there are two sides to paradigms—they are represented by **theory** and by **models of success**. These elements work together to provide justification for your specific course of action.

While each of you will have to explore research in a way unique to your topic, all of you should strive to show that you are not merely asserting your approach to the problem based on opinion, politics, or personal view, but that there is a consensus of opinion or a well-documented trend or development that supports your idea. You will want to discuss theories that form the basis for your assumption that the plan you have in mind will be effective—offering evidence and authority to show that your plan is responding to a body of knowledge in a particular field. If you are planning experimental work that grows out of a well-established scientific paradigm, you should review the tradition of work in the field that you are building on in your research. You will also discuss examples of similar or related projects you are using as models, focusing on the procedures and plans that worked in those instances and emphasizing the positive results achieved. Remember that the main purpose of the research is to justify your plan of action. Thus, if you plan to educate people about a specific environmental issue, you will likely want to focus more on an effective way (or paradigm) of educating people than you will on that environmental issue (though you will need research on that as well).

One of the purposes of the literature review is to establish your authority, which will stand or fall based on the quality of the research you cite. By demonstrating your command over recognized or paradigmatic research, you show that you have the knowledge and expertise to make valid recommendations. You should strive to find the most useful and authoritative research whenever possible, and you should discuss published research (ideally, research that has been subjected to peer review). Many projects will, however, call for a wide range of research sources, including articles, books, Internet sources, published government statistics, interviews, surveys, field studies, calculations, and experimental results. You should do your best to evaluate sources and use only the most solid in building your literature review. To use low-quality materials in constructing your paper is equivalent to using low-quality materials in building a house, and your product will be evaluated and graded (or condemned) accordingly.

8. Plan (or Procedures)

The plan should be as specific as possible and should follow logically from your research. How it is presented will depend upon the specific project you have in mind. If you are proposing a workplace project, you might focus on how your idea will be implemented (perhaps providing a flowchart or time line). If you are proposing to do an experiment, you should lay out the specific procedures you will use. If you are building something, you will want to describe how it will be built and provide diagrams. You might wish to reference research to support the specific choices you are making, though the research section should provide the bulk of your rationale.

9. Budget

The budget should list everything you will need for your project, from salaries to supplies. Some items may require explanation, which you should supply here as well. You should arrange the cost of your budget items in aligned accountant's columns to make your addition clear.

10. Discussion (or Evaluation Plan)

Generally your paper should conclude by summing up your project and making a final pitch for its value. If you are proposing a project whose results can be tested in some way, then you should also offer an evaluation plan.

11. References

This section should list all sources of information cited in your paper in alphabetical order. The bibliography should be prepared according to APA style, covered in the discussion about the APA style guide in Chapter Four. For those who want extra guidance, you might consult *The APA Handbook*, which is available in the reference section of any campus library.

12. Visual Aids (or Figures)

You should have at least three graphic aids that are visual representations of numerical information. These might include graphs, tables, charts, or maps. In addition to these three, you may include drawings, photographs, flowcharts, maps, organization charts, Gantt charts, timelines, diagrams, or floor plans. Each visual graphic aid should be numbered (e.g., Figure 1, Figure 2, etc.) and should have a title. If the graphic is based on information from a source, then you should have a citation line at the bottom (i.e., Source: Alvarez 26). If you can incorporate your graphics into the body of the paper, do so. If you cannot incorporate your graphics, then include them at the end in an appendix or interpaginate them directly following the first reference to them.

13. Appendix (optional)

If you have other information that doesn't exactly fit into your text, you could include it as an appendix (which is literally appended to the end of your document). For example, if you cite a map or chart which is too big to be incorporated into the body of your text, you could label it as Appendix A. Be sure to list it under Appendices in your table of contents, and refer to it in the text (i.e., See Appendix A, p. 20).

Sample Papers

What follows are typical samples of student work, presented for illustrative purposes. This section is included to generate discussion, provide the opportunity for objective critique, and facilitate practicing of contentious peer review. These final proposals are not provided to represent a particular grade or to distinguish between passing and failing work. As with all representative examples, these papers have a variety of strong and weak moments. We encourage you to utilize the course grading criteria to attempt to situate these papers, with the guidance of your instructor. Hopefully, while you are revising your work, this experience will help you to identify the usual moments of achievement, as well as areas that would benefit from improvement.

Proposal #1: Nutritional Program for Adolescents

Jane Hamilton
Rutgers, The State University of New Jersey
341 CPO Way
New Brunswick, NJ 08901
May 16, 2009

Mr. Patrick Spagnoletti
Superintendent of Schools
510 Chestnut Street
Roselle Park, NJ 07204

Dear Mr. Spagnoletti:

The attached proposal is the full and detailed version of my oral presentation which you recently attended. I am sending you the full version to show the importance of my proposal and give you the resources I obtained which prove there is a need for my proposal. I strongly believe this project is of interest to you because you care deeply for the children of Roselle Park not only on a personal level, as the Superintendent of the school district as well. This proposal will address your concern for the obesity epidemic, and it will offer a plan to mediate the epidemic as a whole. Your fears and concerns are understood, and my course of action to remedy obesity will help tackle those uncertainties.

As you know obesity is an epidemic that is growing at an increasing rate in today's society. There are various health risks associated with this disease, most of which are terminal. The scary fact of this epidemic is that it is preventable. What is even more disturbing is the large population of adolescents affected by obesity. Whenever there is a problem, there is most certainly a plan to solve it. I have an insightful plan which addresses the problem of adolescent obesity in today's society, starting with children living in the small borough of Roselle Park, and concluding with the institution of educational activities to begin mediating this problem. Since you have control over the budget, I am hopeful that you will support my conclusion regarding the plan to control adolescent obesity, and give your fiscal support through budget funding to accomplish this project.

Overall it is easy to see the many reasons why my proposal would be beneficial. Children all over the nation are plagued with obesity as a national epidemic. On the road to change, one must start somewhere; why not begin in Roselle Park? If you have any questions or concerns feel free to contact me. My telephone number is (908) 538–6678 and my email address is jham@eden.rutgers .edu. Email is the best way to get in touch with me because of my busy schedule, but I am available to speak on the phone any time before 5:30pm on Mondays and any time between 2:00pm and 5:30pm on Wednesdays. Do not hesitate to call or email me if you have any further questions, comments, or concerns.

Sincerely,

Jane Hamilton

A Plan to Remedy a Worldwide Atrocity: Tackling Adolescent Obesity in Roselle Park

Submitted by:
Jane Hamilton

Submitted to:
Mr. Patrick Spagnoletti
Superintendent of Schools
510 Chestnut Street
Roselle Park, NJ 07204

Date: May 16, 2009
Scientific and Technical Writing

Abstract

In America, 58 million people are overweight, 40 million people are obese, and 3 million people are morbidly obese. Of that population, 11% of children and adolescents are overweight which has increased from 5% in the 60's and 70's. More than one in five children and adolescents ages 6 through 17 are also overweight. The problem stems from improper diet and incredibly low activity levels throughout the United States. Seventy-eight percent of Americans are not meeting basic activity level recommendations. Much of the problem has to do with the fact that people are told false information about what is or is not good for them. They make poor choices based on an artificial foundation of knowledge in the area of nutrition. These poor choices trickle down to their children who also make poor choices. The cycle continues, as does the increasing rate of obesity in America.

There are programs currently installed to help mediate this global epidemic. The Nutrition and Physical Activity Program to Prevent Obesity and Other Chronic Diseases (NPAO) is based on a cooperative agreement between the Centers for Disease Control, Prevention's Division of Nutrition, Physical Activity and Obesity (DNPAO), and 28 state health departments. The program was established in fiscal year 1999 to prevent and control obesity and other chronic diseases by supporting states in developing and implementing nutrition and physical activity interventions. States funded by NPAO work to prevent and control obesity and other chronic diseases through these strategies: balancing caloric intake and expenditure, increasing physical activity, increasing consumption of fruits and vegetables, decreasing TV-viewing time, and increasing breastfeeding. The problem has to stop somewhere. It should start with increased awareness and increased participation in fun and invigorating activities, targeting youth while they are impressionable and still have time to make the right decisions in the area of nutrition. This will be possible through instituting programs in the elementary schools of Roselle Park which will incorporate education on nutrition as well as physical activity and wellness.

Table of Contents

Table of Figures

Introduction

What Is Obesity?

Obesity is defined as a condition in which the natural energy reserve, stored in the fatty tissue of humans and other mammals, is increased to a point where it is associated with certain health conditions or increased mortality. In essence, obesity is too much body fat which can ultimately lead to death. There are many factors that are directly related to obesity; some are within a person's control, while others are not. These factors include unhealthy eating patterns, poor calorie content of food, decrease in physical activity, genetic factors, environmental factors, psychological factors, and cultural factors (CDC, 2007).

Obesity has become a huge issue in today's society. One of the most frightening facts of the global epidemic is that it is extremely prevalent among adolescents. Two studies conducted by the Centers for Disease Control and Prevention in a 1976–1980 and 2003–2004 surveys show that in children aged 2–5 years, the prevalence of overweight kids increased from 5.0% to 13.9%; for those aged 6–11 years, prevalence increased from 6.5% to 18.8%; and for those aged 12–19 years, prevalence increased from 5.0% to 17.4% (CDC, 2007).

Figure 1:
Trends in Child and Adolescent Overweight

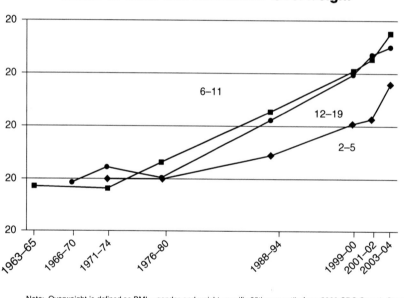

Note: Overweight is defined as BMI>=gender-and weight-specific 95th percentile from 2000 CDC Growth Charts.

***Source:* Centers for Disease Control and Prevention 2007**

These statistics are extraordinarily alarming. A near three times increase of a factor that can be controlled and amended is completely unacceptable. Not only are these statistics alarming, but the correlating health conditions are frightening as well.

There are large array of health conditions which are directly related to obesity. Children can face a future laden with hypertension, dyslipidemia, Type 2 diabetes, coronary heart disease, stroke, gallbladder disease, osteoarthritis, sleep apnea and respiratory problems, and even some cancers-endometrial, breast, and colon (CDC, 2007). These negative effects are incredibly detrimental and horrific. Another alarming fact is that overweight adolescents have a 70% chance of becoming overweight or obese adults (Health and Human services, 2007). It is heartbreaking to know that such a large percent of today's youth will suffer from such horrific consequences directly related to their poor and uninformed decisions regarding their current food consumption and its direct relation to their overall health. This statistic shows that poor adolescent habits in relation to nutrition, health, and an active lifestyle carry over into adulthood.

According to the surgeon general, obesity in children and adolescents is generally caused by lack of physical activity, unhealthy eating patterns, or a combination of the two, with genetics and lifestyle factors both playing important roles in determining a child's weight. Our society has become incredibly sedentary. Television, computers, and video games contribute to children's inactive lifestyles. 43% of adolescents watch more than 2 hours of television each day. Children, especially girls, become less active as they move through adolescence (Health and Human services, 2007). One can see that the main problems are rooted in unhealthy choices of food consumption and lack of physical activity.

A Time for Change

In an interview I conducted on February 20th, 2009 with Kevin Carroll, Physical Education teacher at the elementary schools in Roselle Park, his insight regarding obesity supported the conclusions of the surgeon's general. Mr. Carroll has been involved with Physical Education for 30 years and has seen trends reflecting an increase in more sedated behavior and poor eating habits among his students. When I asked him his opinion on why obesity is such a huge issue in today's society, he offered some insight stating the following:

> "I think the diet of kids is to blame. People are busy working and don't have time to cook, its easier to pick something up from McDonalds. It's easier for them to do that, than prepare a meal. Computer games and video games are to blame too. Also, I think people are scared to let kids just play on their own because of all the creeps out there. But I think diet is a big thing. Fast food is relied on a lot, and personally, children have become more sedated in general. They would rather watch TV and play computer games in general."

Survey Results

It is evident that obesity is a global issue. Considering this, the question is, how does this issue affect Roselle Park, NJ? With permission from Sherman School Principal Faria and the help of Mr. Kevin Carroll, I conducted a general survey with the children grades 3-5. The survey consisted of fifteen total questions [See Appendix pg. 14]. Nine of the questions were simple, short answer questions and the remaining six were "circle the answer that fits you best," both directly pertaining to various aspects of nutrition and lifestyle habits. The data I collected was shocking and showed, as predicted, the trends among the adolescents of Roselle Park mimic those of other adolescents. Of those surveyed, 18 were 5th graders, 34 were 4th graders, and 22 were 3rd graders totaling 74 children in all.

When asked, "What I like to eat is/are:" fruits and vegetables was the highest response accounting for 37% of the children. However, the alarming fact relating to this question is that fast food was the second highest response and accounted for 36% of the children's responses, which is only 1% shy of the highest response. Anything sweet came in next with 12%, followed by junk food with 7%. The combination of anything sweet/fast food and all food were last both with 4%.

Figure 2:
What I Like to Eat Is/Are

Source: **Fun With Food Survey**

The most alarming data came from the question "How many times a week do you eat a dinner that is fast food?" 52% of the children surveyed said they eat fast food for diner 1-4x a week. 26% said they eat fast food for dinner 1-2x a month and 22% said they never eat a dinner that is fast food.

Figure 3:
How Many Times a Week Do You Eat a Dinner That Is Fast Food?

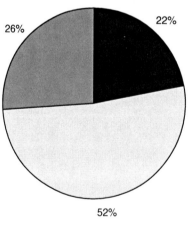

Source: **Fun With Food Survey**

As you can see from the figure below, more than half of the children surveyed are eating a dinner which is fast food 1-4 times a week. This is an important piece of information due to the fact that the fast food industry is a leading contributor to the obesity epidemic plaguing Americans. Fast food chains make it convenient to run in and run out and purchase a "delicious" and "hearty" meal in less than five minutes. Unfortunately, many Americans do not realize the health risk associated with this purchase. Unhealthy eating and the consumption of a very high calorie amount is a very big reason why fast food is such a contributor to obesity.

Summary of Obesity

A changing environment has broadened food options and eating habits. Grocery stores stock their shelves with a greater selection of products. Pre-packaged foods, fast food restaurants, and soft drinks are also more accessible. While such foods are fast and convenient, they also tend to be high in fat, sugar, and calories. Choosing many foods from these areas may contribute to an excessive calorie intake. Some foods are marketed as healthy, low fat, or fat-free, but may contain more calories than the fat containing food they are designed to replace. It is important to read food labels for nutritional information and to eat in moderation. Portion size has also increased. People may be eating more during a meal or snack because of larger portion sizes. This results in increased calorie consumption. If the body does not burn off the extra calories consumed from larger portions, fast food, or soft drinks, weight gain can occur (CDC, 2007).

This information is especially pertinent because children are consuming an abundance of calories, which combined with their sedentary behavior, is directly contributing to obesity among adolescents. All of this information is frightening and overwhelming and can make it is easy to see how one could easily take a path that leads towards obesity. These statistics are not just numbers; they are people, children, with bright and promising futures. The fact of the matter is that this disease, which our nation is undoubtedly plagued with, is entirely preventable. The plan for Roselle Park is to tackle the causes of obesity head on, and help the children of Roselle Park build a strong foundation on which they can make healthy and wise decisions regarding food consumption and vigorous activities in their lives.

Literature Review

Healthier Generation

Everyday in the United States 53 million people go to school to work or to learn. That is about one in five Americans who will spend around 30 hours of their week in a school. As a result, schools are a powerful way to shape the health, education, and well-being of Americans. The Alliance for a Healthier Generation has developed criteria to help schools identify concrete actions they can take to establish a healthier school environment and they reward schools which have met the criteria established (Healthier Generation, 2007).

The basic framework is centered on the following components:

- Systems and policy
- School meals
- Competitive foods
- Health education
- Physical education
- Physical activity opportunities
- After school programs
- Staff wellness programs

Social-Ecological Model

Elsewhere, a Social-Ecological Model has been successfully accomplished. The model starts with individuals by changing everyday behaviors that relate to eating and physical activity as well as changing people's knowledge, attitudes, and beliefs. It creates organizations which help make better choices about healthful eating and physical activity as well as make changes to organization policies and environments by providing health information. The organizations in turn trickle down to the community and eventually reach society as a whole (HHS, 2007).

**Figure 4:
Social-Ecological Model**

Source: **Health and Human Services**

NPAO

The Nutrition and Physical Activity Program to Prevent Obesity and Other Chronic Diseases (NPAO) is based on a cooperative agreement between the Centers for Disease Control and Prevention's Division of Nutrition, Physical Activity and Obesity (DNPAO) and 28 state health departments. The program was established in fiscal year 1999 to prevent and control obesity and other chronic diseases by supporting states in developing and implementing nutrition and physical activity interventions (CDC 2007).

States funded by NPAO work to prevent and control obesity through these strategies:

- Balancing caloric intake and expenditure

- Increasing physical activity

- Increasing consumption of fruits and vegetables

- Decreasing TV-viewing time

NPAO states that it is vital to teach skills needed to make individual behavior changes related to nutrition, physical activity, and healthful weight, and provide opportunities to practice these skills (CDC 2007). The key goal of the NPAO is teaching skills. Correspondingly, it is important to create supportive environments, which make healthful lifestyle options more accessible, affordable, and safe. The program provides resources, training, and mini-grant funding for schools to make changes to the school environment, including more healthful food choices. It gives them the ability to establish programs in communities to increase physical activity and/or reduce caloric intake through healthful eating habits (CDC 2007).

These programs are currently being run in 28 states across the U.S. One state in particular is Massachusetts. Massachusetts integrated nutrition and physical activity into existing core subjects and assessed the nutrition and physical activity policies. They created before-school and after-school nutrition and physical activity programs while launching school-wide campaigns to promote the 5-2-1 message:

- Eat at least five fruits and vegetables per day
- Watch two hours or less of TV per day
- Engage in at least one hour of physical activity per day

Program ENERGY

A groundbreaking program in Colorado is tackling the problem of obesity head on. It focuses on teaching children to incorporate the elements of a healthy lifestyle which in turn will help reduce the risk of chronic diseases. Two components of the intervention are:

1. The use of scientists in the classroom to lead hands on challenging and fun lessons providing information about healthy lifestyle, how the body works, and science.

2. Educating on maintaining a healthy weight, being physically active, and eating a healthy diet.

Each week the program provided interactive, hand-on, fun educational science enrichment activities. Lessons included nutrition/science games, energy balance, and healthy snacks. Year 1 pre and post tests showed meaningful increases in health and science knowledge as well as increased appreciation for healthy food selection and physical activity. It was scientifically concluded that program ENERGY has changed the health behaviors of elementary school students.

Studies were conducted, and results showed that children with iron deficiencies sufficient to cause anemia are at a disadvantage academically, unless they receive iron therapy. Improper nutrient intake can lead to anemia and other deficiencies. The study also showed that food insufficiency is a serious problem affecting children's ability to learn; and that offering a healthy breakfast is an effective measure to improve academic performance and cognitive functioning among undernourished populations (Taras, 2005). This proves that educating children on nutrition will increase academic performance.

All of the information above proves just how vital education is to impacting ones choices. Educating the children of Roselle Park will help them make better choices in the present which will have long-lasting benefits for their futures.

Plan

Main Objective

My plan and course of action is heavily based upon instituting educational programs at the Roselle Park elementary schools. I will conduct fun, educational activities about health and nutrition. The activities will range from teaching the children why the body needs food to why certain foods are better than others, and how to make good decisions regarding food. This reflects information from scholars who state promoting healthier eating habits is a method that could be used to prevent obesity and overweight in children (Motycka and Inge, 2007).

Since exercise and activity are incredibly important to fighting the increasing rate of obesity, I will include activities and suggestions that they can do rather than the sedentary activities most of today's youth participate in on a regular basis. Motycka and Inge also state that promoting physical activity at school can assist in preventing this epidemic from expanding. They conclude that it is important that children change the way they behave every day. Such as walking to school, taking the stairs instead of the elevator, and even how often they walk around every day as opposed to sitting down to study or watch television, all of which have a great impact on energy expenditure.

Setting Up Educational Program

My programs will run after school so as to not interfere with current curriculum. The current after care programs already running at the elementary schools would be a perfect place to run these educational programs, especially since most of the children who participate in after care are the children of parents who lead particularly busy lifestyles and are at a greater risk to have poor eating habits based on fast food trends and other lifestyle factors.

The frequency of the program is speculative; however, the aim is to run the program two times a week throughout the entire academic year. Each program information session will be roughly one hour long and will pertain to a different topic. The information session will be interactive and incorporate visual aides and hands on tools to help educate the children on important concepts like portion control and proper nutrient intake decisions.

Following the informational session will be a physical session which will be roughly one hour long and include an invigorating activity which will vary week to week. Since the current after care program in place is already a safe environment for the children, the concern for a safe place to exercise is met.

Information covered in the information session will include:

- The science of nutrition
- Assessing nutritional status and guidelines for dietary planning
- Chemical and biological aspects of nutrition
- Nutrition physiology: digestion, absorption, circulation, and excretion
- Carbohydrates
- Protein
- Lipids
- Energy metabolism
- Energy balance and body weight regulation
- Water soluble vitamins
- Fat soluble vitamins
- Trace minerals
- Portion control
- Choosing nutrient dense foods

The overall basis of both information and physical sessions is to encourage general lifestyle changes. It is important to incorporate various aspects of the child's lifestyle into the sessions so that there can be an all around change in the way the child behaves.

Children have the ability to make proper choices regarding their eating habits. What I find they lack are the tools to aid them in making these proper decisions. The knowledge and strength they will gain from my instructional workshops will greatly increase their awareness and ability to make conscientious decisions about nutrition and ultimately their overall health.

Budget

Rationale

The following budget is based upon the conditions of 36 weeks in a given academic year with the program meeting 2 times a week totaling 72 2.5 hour program sessions.

1. Salaries...$5,400.00

2. Supplies...$4,320.00

3. Space...Free

4. Food...$9,720.00

   TOTAL $19,440.00

Justification

1. Salaries

   - With 72 total sessions and roughly 2.5 hours per session earning $10.00 an hour including all 3 elementary schools the total comes to $5,400.00

2. Supplies

   - Roughly $20.00 a session for expenses from visual aids and interactive portions of the session including all 3 elementary schools the total comes to $4,320.00

3. Space

   - The current site for the programs is the after school after care program which is already set up at each elementary school which will be of no cost to run

4. Food

   - A sample snack on any given day will vary but an example of the expense will stem from the following possible snack:
   - Vegetable platter...$25.00
   - Granola Bars...$15.00
   - Low Fat milk..$5.00
   - TOTAL $45.00
   - For 72 total sessions including all 3 elementary schools the grand total comes to $9720.00

Discussion

Obesity is a huge issue in society. There is an alarming amount of data showing a direct causal relationship between food consumption and obesity. It is important to take this data into consideration when trying to remedy the problem. If children are properly educated on nutrition they will have the tools necessary to make educated decisions in regards to food intake.

There is a vast array of programs set up throughout the nation to help combat the obesity epidemic. All of these programs instituted embody some aspect of education, physical activity, and student involvement. By educating children on food and increasing their physical activity levels, obesity has been proven to show signs of reversal.

Ultimately, my plan of action is to implement educational and physical programs during the after school after care programs in the Roselle Park elementary schools. The information portion of the sessions will be fun and interactive and include information on food intake, nutrient density, portion control, and overall lifestyle changes. The physical portion of my program will incorporate fun and exciting physical activities which will be fresh and interesting week to week.

Evaluating the progress of my program is something that I am incredibly interested in pursuing. Since education is vital to the reversal of the epidemic and is one of the prominent aspects of my proposal, bi weekly assessments will be given on topics learned from the week prior. The children will be assessed on conceptual understanding of topics and general application of the knowledge they have acquired. The success of the program can also be measured in an increase in interest in physical activity overall, which will be measured by enthusiasm for participation in gym and other physical activities.

Overall, it is easy to see the many reasons as to why my proposal would be beneficial. Children all over the nation are plagued with obesity as a national epidemic. On the road to change, one must start somewhere; why not begin in Roselle Park? I will close with one of my favorite quotes: *The only limit to our realization of tomorrow will be our doubts of today*"-FDR. I ask you not to doubt my proposed plan of action; through my careful and concise research I am certain it will be exceedingly effective in combating obesity in Roselle Park. With budget funding to help propel this project, I strongly feel we can combat the proposed problem, all I ask is your help in the fight.

References

Alliance for a healthier generation. Retrieved May 7, 2009, from the website: http://www.healthiergeneration.org/ default.aspx

Beyond boundaries. Retrieved May 5, 2009 from the website: http://www.tufts.edu/development/news/2007/ shapeup.html

Boyle, M.A. & Long, S. (2007). *Personal nutrition, sixth edition.* California: Thomson Wadsworth.

Centers for disease control and prevention. Retrieved May 7, 2009, from the website: http://www.cdc.gov/ nccdphp/dnpa/obesity/

McGuire, M & Beerman, K.A. (2007). *Nutritional sciences: from fundamentals to food.* California: Thomson Wadsworth.

Motycka, Carol and Inge, L.D. (2007). The growing problem of childhood obesity. *Drug topics,* 151(17), 33–42.

Taras, Howard (August 2005). Nutrition and student performance at school. *The journal of school health,* 75(6), 199–213.

The American physiological society. May 1, 2009, from the website: http://www.the-aps.org/press/conference/ eb03/9.htm

The finance project. Retrieved May 1, 2009, from the website: www.financeproject.org/Publications/ obesityprevention.pdf

The obesity society. Retrieved May 7, 2009, from the website: http://www.obesity.org/

United States department of health and human services. Retrieved May 1, 2009, from the website: http://www.surgeongeneral.gov/topics/obesity/calltoaction/fact_adolescents.htm

Wardlaw, G.M., Hampl, J.S. & DiSilvestro, R.A. (2004). *Perspectives in nutrition.* New York: McGraw-Hill.

Appendix

Fun with Food and Activity Survey

Directions: Circle the one answer you think fits you best. NO answer is a wrong answer.

1. The reason I need to eat is because:
   - I need energy to play and learn
   - It makes me happy
   - It's fun when I get bored

2. What I like to eat is/are:
   - Anything sweet (deserts/candy)
   - Fruits and vegetables
   - Fast food (McDonalds/Burger King)
   - Junk food (Chips/Pretzels)

3. I like to eat food:
   - With my friends
   - Alone
   - While I'm watching T.V.
   - With my family

4. After school I:
   - Like to watch T.V.
   - Like to go outside and play
   - Like to play video games
   - Do homework

5. I know that I'm full:
   - When my stomach hurts
   - When there is no food left on my plate
   - I'm never full I could eat all the time
   - When I'm not hungry anymore

6. The times I eat are:
   - Whenever I get hungry
   - Whenever there is food around
   - Breakfast Lunch and Dinner
   - Whenever I am told I should

Directions: Write the answer to the question in the empty space. NO ▓ answer is a wrong answer.

1. How do you feel about the way you look?

2. How many times a week do you eat dinner that is fast food?

3. How many times a week do you eat dinner that is cooked at home?

4. What does the word exercise mean to you?

5. What does the word calorie mean to you?

6. What is your idea of the perfect meal?

7. What is your least favorite food?

8. What is your favorite food?

9. What food is the best for your growing body?

Proposal #4: Childhood Lead Exposure Awareness

Jill Wu
651 CPO Way
Rutgers University
New Brunswick, NJ 08901

May 13, 2009

Rochelle D. Williams-Evans, RN, MS
Director of Health & Human Services
East Orange Health Department
143 New Street, East Orange, NJ 07017-4918

Dear Ms. Rochelle D. Williams-Evans,

Thank you for attending the presentation of my proposal: Implementing a Childhood Lead Exposure Education and Awareness Program in East Orange, New Jersey. As stated in my presentation, childhood lead exposure is an ongoing health issue that is devastating the lives of many families within the community of East Orange. It is in the interest of the Health and Human Services Department to work towards maintaining the safety, health and well being of men, women and children. Lead poisoning poses an enormous threat to the health, education and personal outcome of a growing child. Due to the lack of parent knowledge of the overall dangers of lead poisoning and the importance of testing children for lead exposure, many children are suffering the consequences. Since this town contains some of the highest percentages of children with blood lead levels over the accepted state level within New Jersey, it is crucial that this issue is properly solved. By addressing this public issue, Health and Human Services can continue to maintain the health and safety of everyone within East Orange.

To address this issue, I propose a plan to educate parents and expecting parents on the need to have children screened for lead in blood, health problems associated with lead poisoning, and several other significant facts about childhood lead exposure. My program will involve the distribution of brochures at the local medical clinic, as well as neighborhood pediatric and OB/GYN offices. I will also display public advertisements in the community newspaper. Lead education and awareness programs performed in the past have been proven effective in increasing the amount of childhood lead screenings and spreading awareness of the dangers of the lead poisoning. Specific guidelines are thoroughly included within this proposal package regarding this new service.

If you have any questions regarding this proposal or would like to meet with me personally, please feel free to contact me via email at jwu@eden.rutgers.edu, or by phone at 732-463-5572. Thank you so much for your time and consideration and I look forward to hearing from you.

Sincerely,

Jill Wu

Implementing a Childhood Lead Exposure Education and Awareness Program in East Orange, New Jersey

Submitted by:

Jill Wu

Submitted to:

Rochelle D. Williams-Evans, RN, MS
Director of Health & Human Services
East Orange Health Department
143 New Street, East Orange, NJ 07017-4918

Submitted on:

May 13, 2009

Scientific and Technical Writing
Course Number: 355:302:09

Abstract

Childhood lead poisoning is a serious condition that can have a negative impact on the life of a growing child. This severe health issue is an ongoing problem for the town of East Orange because many people are unaware of the importance of testing children for lead poisoning and the dangers associated with lead exposure. The proposal describes a detailed plan to help lessen or even prevent this prevailing concern. The paper argues that an education and awareness program, where brochures and advertisements are placed throughout the community, is the most effective strategy for this problem. This proposal is modeled after the lead education programs provided by the Division of Medical Assistance and Health Services and the American Civil Liberties Union Foundation, the Massachusetts Childhood Lead Poisoning Prevention Program, and the Environmental Protection Agency.

Brochures, containing thorough information about childhood lead exposure and the need to have children tested, will be available for parents and expecting parents to read at various pediatric and OB/GYN offices throughout East Orange. Also, advertisements will be available for everyone to read in the local newspaper, *The East Orange Record*. Informing the public is predicted to increase the number of child lead screenings and ultimately prevent childhood lead exposure.

Table of Contents

Table of Figures

Introduction

The county of East Orange faces an enormous health problem: childhood lead exposure. It is in the best interest of the Health and Human Services Department to address this prevalent issue before another child's life is jeopardized by lead poisoning. Studies have shown that if lead poisoning is not detected early in children, high blood lead levels can cause numerous health problems such as slowed growth, hearing problems, and brain damage (*Lead in paint, dust, and soil*, 2007). These health issues, along with many more, can devastate a child's education and potential to live a productive and vigorous life. This serious issue needs to be addressed and properly solved before childhood lead poisoning reaches a level that is detrimental and irreversible. I would like to propose a solution that benefits both the East Orange Health Department and the families within the community.

Childhood Lead Poisoning in East Orange, New Jersey

Childhood lead exposure is an ongoing health problem that has the potential to devastate the life of any living child. Within the state of New Jersey, every county contains children suffering from lead poisoning (Chen, 2006). The Department of Health and Senior Services' fiscal year 2003 Report compared the blood lead levels of children in New Jersey. Out of all the municipalities in New Jersey, East Orange City contains the greatest percentage of children tested who had blood lead levels greater than the states' accepted blood lead "level of concern," of ten micrograms/ deciliter (Chen, 2006).

Figure 1:
Percentage of Children Tested with Blood Lead Levels
Greater than 10 Micrograms/Deciliter

East Orange	9.9%
Irvington Township	9.4%
Newark City	8.1%
Trenton City	8.1%
Paterson City	6.5%
Plainfield City	4.4%
New Brunswick City	4.3%
Montclair Township	4.3%
Atlantic City	3.9%
Passaic City	3.8%
Elizabeth City	3.2%
Hamilton Township	3.2%
Lakewood Township	3.1%
Jersey City	3.0%
West Orange Township	2.8%

Source: **Chen, 2006**

In comparison to other municipalities in New Jersey, it is evident that lead poisoning is a serious health issue in East Orange. After talking about two young children in East Orange that have severe health problems as a result of lead paint exposure, Russel Ben-Ali and Judy Peet (2005) state, "The Washington children are among thousands of youngsters throughout New Jersey caught in a toxic trap" (pp. 5–6). Although the statewide blood lead "level of concern," implies that any child under 10 micrograms per deciliter of lead in blood is not at risk of any dangerous health problems, "The Centers for Disease and Control make it clear that there is no known safe level of lead toxicity" (Chen, 2006). Even after it has been proven by scientists that any amount of lead paint in blood can pose an immediate threat to the life of a growing child, East Orange still has "150 children with lead levels between 15 and 20 micrograms" (Ben-Ali, 2005, pp. 5–6).

Health Problems Associated with Childhood Lead Paint Exposure

There are various health and learning problems that are likely to occur when a child has lead poisoning. The consequences of lead paint exposure can ultimately impair the life of a growing child. "It is well established that children under age 6 are especially vulnerable" (Jacobs & Nevin, 2006, p. 2). A special report from the Association for Children of New Jersey states that this is due to that fact that children's bodies, under the age of six, "absorb lead more easily and at a greater rate than adults. If untreated, lead poisoning can cause behavioral difficulties, learning disabilities, retard development, seizures, coma, severe brain and kidney damage and even death" (*Eliminate Childhood Lead Poisoning,* 2007*).* Studies show that, "lead paint poses health problems at any age but its potential to alter cell structure and chemistry of developing brains can devastate young children" (Ben-Ali, 2005, p. 6).

In New Jersey, thousands of children suffer from the effects of lead paint exposure. There is an enormous list of permanent health effects from lead poisoning, such as mental retardation, reduced IQ, learning and reading disabilities, hyperactivity, and developmental problems in most bodily organs, particularly the central nervous system, red blood cells and the kidneys (Chen, 2006). Children six years old and under, and fetuses, will suffer the most from lead exposure because, "they are particularly vulnerable because at that age, their brain and central nervous system are still forming" (*Health effects on children,* 2004, 1). Although the symptoms of lead poisoning in children are sometimes difficult to detect, a brochure released from the United States Environmental Protection Agency states that, "children with high levels of lead may complain of headaches or stomachaches, or may become irritable and tired . . . The only sure way to know if a child has too much lead in his or her body is with a simple blood test" (2004). These health problems obtained early in a child's life can seriously alter his or her education, daily lifestyle and outcome in society.

How Children Are Exposed to Lead Paint

Every year in East Orange City, men and women come together to raise children of their own, in hopes that their family will be healthy and safe. The only problem is many of these families are living in housing units that contain dangerous amounts of lead that are hazardous to the health of themselves and their children. Across the state of New Jersey, especially in East Orange, there are hundreds of occupied buildings that still contain lead. The federal government did not pass a law that banned lead-based paint from housing units until 1978; leaving hundreds and thousands of buildings, still today, contaminated with lead (*Lead in paint, dust, and soil,* 2007*).* Since most homes built before 1978 used lead-based paint, many families within the state of New Jersey are living in older buildings that may contain this toxic substance. "Despite the laudable efforts by various State agencies in reducing the incidence of childhood lead poisoning in New Jersey, at least two-thirds of New Jersey housing still contains lead-based paint and thousands of New Jersey's children, especially those who are the most impoverished and powerless, still suffer" (Chen, 2006). It is evident that even after State agencies' efforts, childhood lead exposure is still a current issue in East Orange, New Jersey.

Children can become poisoned from lead from a variety of sources. Serious hazards include paint chips or paint dust on windows, windowsills, doors, doorframes, stairs, banisters, railings, fences and porches (*Lead in paint, dust, and soil,* 2007*).* Since children do not know any better than to put their fingers and/or objects in their mouths, they are a lot more susceptible to ingest these poisonous materials and suffer the health problems.

Childhood Lead Exposure and Parent Involvement

A lead blood test is the only way to insure that a child has or has not been exposed to lead. The Lead Poisoning Abatement and Control Act requires, "Health care providers to screen all children for elevated blood lead levels at both 12 and 24 months of age. Older children, up to age six, are also to be tested if they have not previously been tested, or are assessed to be high risk" (*Eliminate childhood lead poisoning,* 2007, p. 1). It is essential that children undergo a lead blood test, or a "lead screening", especially if they are living in a building that was built before 1978. Although the law does not require testing children over the age of six, tests should still be run upon previously elevated test results or other risk factors (*Eliminate childhood lead* poisoning, 2007, p. 1). Hundreds of undiagnosed children suffer each year because many parents do not take their sons and/or daughters to get tested for lead exposure. "In FY 2002, only 40% of the estimated 222,800 children between 6 and 29 months of age received a lead screen" (*Eliminate childhood lead poisoning,* 2007, p. 1). It is crucial that parents understand the importance of getting their child's blood tested for lead content.

A parent's lack of knowledge of the health problems associated with childhood lead poisoning can be extremely detrimental to the life of a young child. In Michigan, where childhood lead poisoning is also a prevailing issue, Representative Carl Williams states, "Too many children are not being tested. There is a lack of education; most people don't realize that lead is everywhere. It's not just in old homes in the inner city and old farm-houses" (Schneider, 2004). To prove the lack of parent knowledge of the dangers associated with lead poisoning in East Orange, New Jersey, I conducted two surveys. After asking thirty people a series of two questions, both surveys concluded that there is not enough education and awareness on lead poisoning.

Figure 2:
Survey #1: How Big of a Problem Do You Think Lead Exposure Is to One's Health?

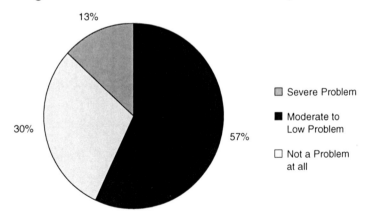

Figure 3:
Survey #2: How Much Have You Heard about the Problem of Lead Exposure throughout Your Lifetime?

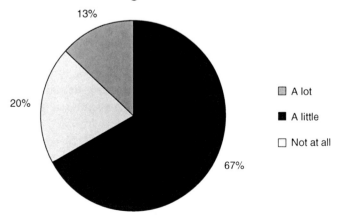

Since 57% of the thirty people I surveyed think that lead poisoning can only pose a moderate threat to one's health and only 13% believe it can be a severe problem, it is evident that East Orange residents are not aware of the health hazards associated with lead poisoning. The second survey shows that people are uninformed and unaware of how serious of an issue lead poisoning is in East Orange, New Jersey. I have therefore concluded that there are not enough educational resources available to the residents of the East Orange community. According to the Centers for Disease Control and Prevention, "increased community-wide awareness can generate broad commitment to improve community resources and political will for primary prevention" (Brown, 2005). Spreading awareness of the severe health issue of childhood lead poisoning can improve parent knowledge and ultimately reduce childhood lead exposure.

Literature Review

Childhood lead poisoning is a widespread health issue that impacts the lives of families across the country. There have been several approaches across the United States to try and put an end to childhood lead poisoning. Most of the attempts to end increasing blood lead levels in children have been conducted through an educational approach. Spreading awareness of the harmful effects of lead poisoning and the need to have children tested has had a positive effect on the overall reduction of childhood lead exposure.

At the College of Nursing in The Ohio State University, Dr. Barbara Polivka (2006) and a team of nurses conducted a survey to determine the more favorable method of receiving lead poisoning prevention information. Results showed that, "brochures and discussions with health care providers were the preferred methods of obtaining lead-poisoning prevention information." Younger parents, especially, preferred to acquire knowledge through billboards, brochures or speaking with someone at the health department. (Polivka, 2006). This study shows that using public advertisements would be the most favorable and influential approach to informing parents about the hazards of lead paint exposure.

Massachusetts Childhood Lead Poisoning Prevention Program

The Massachusetts Childhood Lead Poisoning Prevention Program (MA CLPPP) introduced an educational program that targeted pregnant women in Massachusetts. In order to educate expecting mothers about the dangers of childhood lead exposure and "encourage them to adopt preventative behaviors," the MA CLPPP's project involved the, "development and distribution of bilingual prenatal lead awareness kits packaged in large attractive diaper bags . . . including educational fact sheets and brochures, promotional items, a community resource card, an evaluation card, and a voucher for free lead-safety training for a family member" (Brown, 2005). Targeting pregnant women and expecting parents can help prevent a child from ever becoming exposed to lead and suffering the consequences. Since this program is fairly new to the state of Massachusetts, the effectiveness has not been determined. Another plan, using the same approach as MACLPPP, has been proven very successful in efforts to increase parent knowledge of childhood lead exposure and ultimately increase the amount of lead screenings in children.

American Civil Liberties Union Foundation and the Division of Medical Assistance and Health Services

In addition to East Orange, Camden City and Irvington are two municipalities that contain a substantial number of children with elevated blood lead levels. The Division of Medical Assistance and Health Services ("DMAHS") and the American Civil Liberties Union Foundation ("ACLU") came together to propose a project in Camden City and Irvington. Their plan was to "increase the lead screening of Medicaid-enrolled children under the age of six"(*Eliminate childhood lead poisoning,* 2007, p. 2). The groups' approach was to have trusted health care providers and agencies, such as day care employees, inform parents of the hazards of lead poisoning and the need to get children tested. Sources of education included discussions, videos, and informational packets (*Eliminate childhood lead poisoning,* 2007, p. 2).

This study proved that it is indefinitely possible to mobilize a community, educate and encourage parents to test their children for lead poisoning. Results from the city agency, that overseas day care centers, showed that after more parents were informed, documented lead screenings increased by twenty to thirty percent (*Eliminate childhood lead poisoning,* 2007, pp. 2–3). The pilot proves that, "the child care community and health care providers can play a successful role in eliminating childhood lead poisoning" (*Eliminate childhood lead poisoning,* 2007, p. 3). The success of this plan shows that education is the key to increasing the amount of lead blood tests in children and ultimately decreasing the amount of childhood lead exposure.

Environmental Protection Agency

The Environmental Protection Agency (EPA) released brochures in both 2003 and 2004 containing several different topics involving how to protect you and your family from the hazards of lead. A person that is unaware of the dangers of childhood lead exposure can easily be informed after reading one of these information packets. Examples of the various topics discussed in the brochures include: how children contract lead poisoning, the health problems associated with childhood lead exposure, and the need to have children screened for lead in blood (USEPA, 2004). The EPA also informs parents how exactly lead gets into the body and why "lead is even more dangerous to children under the age of six" (USEPA, 2003). These brochures provide helpful facts in a way that is informative and easy to interpret. By distributing informational packets, similar to these, families of East Orange will be more knowledgeable on the issue of childhood lead exposure. This spread of awareness will not only benefit mothers, fathers and expecting parents, but the lives of children suffering from lead poisoning.

These models are excellent examples of how education and awareness programs can be used to reduce and ultimately prevent childhood lead exposure. If the proper education and awareness is given to men and women before they have children, more and more children can be saved from ever developing the health problems associated with lead poisoning. Preventative action needs to be taken in order to help the lives of children in East Orange, NJ.

Procedures

After extensive research of various techniques used to fight and end childhood lead exposure, I have formulated a plan that is bound to be efficient in solving this severe health issue. Many of the successful models above used educational programs to inform people of the dangers of childhood lead poisoning. My program will involve the use of brochures and advertisements, containing information about the dangers associated with childhood lead exposure and the need to get children tested for lead poisoning. Centers for Disease Control and Prevention state that lead education and awareness programs should include "information about lead poisoning, evaluation and control of lead hazards, home preparation, local lead safety resources and community groups, and screening recommendations" (Brown, 2005). I will be targeting pregnant women, expected parents, and parents with young children.

I will reach out and educate men and women that are unaware of this extreme health issue occurring in their own neighborhood. Steven Marcus of the University of Medicine and Dentistry, a nationally recognized expert on lead poison treatment, states, "that would be the real gold standard- true prevention, not waiting until any kid has elevated lead level" (Ben-Ali, 2005, p. 6). Increasing awareness and improving parent education on this enormous issue can prevent a child from ever becoming exposed to lead.

Since expected parents, pregnant women and/or parents of young children frequently visit their local pediatrician or OB/GYN, the brochures I have designed will be placed at these two types of health care facilities throughout the city of East Orange for adults and/or parents to read. These brochures will be located at nine pediatric offices and six OB/GYN offices throughout the community. In addition to these medical offices, I will also place these informational packets at the local medical clinic. I have chosen to include a medical clinic because many families are not financially able to afford a pediatrician and therefore seek medical assistance at local clinics. It is essential that the entire community is included in this plan and no one is left behind due to an economical disadvantage.

The attached brochure contains:

- The definition of lead poisoning
- How children are exposed to lead
- The various health problems associated with lead poisoning
- The symptoms of childhood lead exposure
- The need to talk to your doctor about testing your children for lead poisoning
- The National Lead Information Center phone number for more information regarding childhood lead exposure

In addition to brochures planted in various health offices, I would like to reach out and inform people of East Orange through public advertisements in the community newspaper, *The East Orange Record* (Local Source, 2007). "Since many young children are not routinely screened as part of their regular physicals or 'well child' visits . . . large numbers of lead burdened children are undiagnosed and untreated" (*Eliminate childhood lead poisoning*, 2007, p. 1). As a result of this knowledge, I will be stressing the need to get children screened for lead exposure in my public advertisements. This public health announcement in newspapers will inform the East Orange community and increase the amount of lead screenings in children. Testing early for lead poisoning can prevent a child from ever developing elevated blood lead levels and suffering the health problems associated with lead exposure.

The knowledge and awareness of the severity of childhood lead poisoning and the absolute need to have children tested is essential to ending the prevailing issue of childhood lead exposure in East Orange, New Jersey. It is crucial that each and every parent and expecting parent is completely informed of this health problem before another young boy or girl's life is damaged due to lead poisoning.

Budget

Brochures:

16 Total Destinations

9 Pediatric Offices, 6 OB/GYN Offices, 1 Medical Clinic

100 brochures a month per destination = $40.00

Year supply of brochures per destination = $480

Total price of brochures $7,680

Deliver Brochures:

Delivered at the beginning of every month by UPS

$70.50 per delivery for 12 months $846

Newspaper Advertisement:

$70 per ad (1 per week) × 52 weeks $3,640

Total Annual Cost $12,166

Discussion

Childhood lead poisoning is a widespread health issue that is destroying the lives of innocent children throughout East Orange each and every day. It is evident that the East Orange community is unaware of the severity of childhood lead exposure and the huge impact it can have on one's life. There is an absolute need for more available resources to educate the residents of this town. Knowledge and awareness of lead poisoning can be improved by using public advertisements and brochures that contain thorough information on the dangers of childhood lead exposure and the need to have children tested. The proper knowledge of expecting parents and current parents of young children on childhood lead exposure can help increase the number of lead blood tests performed, prevent the risk of raising a child in a lead infested environment, and ultimately lower blood lead levels in children. Due to successful lead prevention programs that targeted parent education performed in the past, I feel as though my plan will help the Health & Human Services Department reach their goal of sustaining the well-being of individuals within the town of East Orange. This education and awareness program will not only help hundreds of families within this city, but also strengthen your department and the overall service to your community.

References

Ben-Ali, Russell & Peet, Judy. Jersey children caught in a toxic trap. (2005, December 2). *The Star Ledger.* p. 1, 2.

Brown, Mary Jean. (2005). Building blocks for primary prevention: Protection children from lead-based paint hazards. *Center for Disease Control and Prevention.* Retrieved October 21, 2007, from http://purl.access .gpo.gv/GPO/LP576979.

Chen, Robert K. Brief of Amicus Curiae public advocate of New Jersey. *(2006, April 20). Retrieved May 6, 2009, from http://www.state.nj.us/pubicadvocate/reports/pd fs/Lead _Paint_Amicus_Brief_Final_2_20_06.pdf.*

Staples Custom Printing. (2007). *Retrieved May 8, 2009, from http://www.staples.com/webapp/wcs/stores/ servlet/StaplesCategoryDisplay?storeId=10001&catalogIdentifier=2&identifier=CG1000*

Eliminate Childhood Lead Poisoning: Special Report. (2004, February). Retrieved May 6, 2009, from http://www.kidlaw.org.main.asp?uri=1003&di=320&dt=4x=1.

Health effects on children. (2004, March). Retrieved May 1, 2009, from http://www.nsc.org/issues/lead/ healtheffects.html

Jacobs, D.E. & Nevin, R. (2006). Validation of a 20-year forecast of US children lead poisoning: Updated prospects for 2010. *Environmental Research,* 102, 352–364. Retrieved May 2, 2009, from Environmental Sciences & Pollution Management Index database.

Lead in Paint, Dust, and Soil. (2007, August 2). Retrieved May 2, 2009, from http://www.epa.gov/lead/pubs/ leadinfo.htm#health.

Local Source. (2007, December 8). Retrieved May 8, 2009 from http://www.localsource.com/advertise/

Polivka, B.J. (2006). Needs assessment and intervention strategies to reduce lead-poisoning risk among low-income Ohio toddlers. *Public Health Nursing,* 23, 52–58. Retrieved October 2, 2007, from Environmental Sciences & Pollution Management Index database.

Schneider, Kathryn.(2004). Michigan seeks reduction in lead poisoning, increase in testing. *Stateline, 4.* Retrieved May 3, 2009, from http://www.csg.org/pubs/Documents/slmw-0407Michiganseeks.pdf

United States Environmental Protection Agency: Office of Pollution Prevention and Toxics. (2003). Protect your family from lead in your home [electronic version]. *IRIS.*

United States Environmental Protection Agency: Office of Pollution Prevention and Toxics. (2004). Give your child the chance of a lifetime: Keep your child lead safe [electronic version]. *IRIS.*

UPS Calculate Time and Cost. (2007). Retrieved May 8, 2009 from http://wwwapps.ups.com/calTimeCost?loc=en _US&WT.svl=PNRO_L1

Appendix: The Brochure
"Help Lead the Way Against Lead"

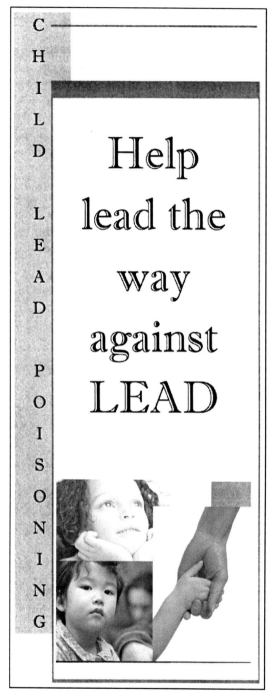

CHILD LEAD POISONING

Help lead the way against LEAD

What is Lead Poisoning?

- Lead poisoning, a major health issue in the United States, is a preventable medical condition that involves elevated blood lead levels

How are Children Exposed to Lead?

- Since most homes built before 1978 used lead-based paint, many families within NJ are living in buildings containing this toxic substance

- Serious hazards include: Paint chips or paint dust on windows, windowsills, doors doorframes, stairs, banisters, railings, fences, and porches

- Since young children are more likely to put their fingers and/or objects in their mouths, they are more susceptible to become exposed to lead

What are the Health Problems Associated with Lead Poisoning?

- Brain, nervous system and kidney damage

- Learning & reading disabilities

- Attention deficit disorder

- Decreased Intelligence

- Hearing damage, speech and behavior problems

Photos © JupiterImages.

What are the symptoms of Childhood Lead Exposure?

- Childhood Lead Exposure is also known as, "The Silent Menace," because there are not many symptoms.

- Some children complain of headaches, stomach pain, fatigue and feeling tired

* The ONLY sure way to know if a child has been exposed to lead is with a SIMPLE BLOOD TEST *

Talk to your child's doctor about having a lead blood test TODAY

... before it is too late

Give your child the life they deserve

For more information on lead poisoning, please contact the National Lead Information Center:

1-800-424-LEAD

(English or Spanish language)

Photos © JupiterImages.

Reader: _____   **Recipient:** _____

■ Final Proposal Workshop I

Please fill out the following form for your partner. Feel free to write comments on the drafts as well.

Cover Letter and Title Page

Does the cover letter . . .

1. directly address the funding source?	_____yes _____no
2. explain why the reader has received this proposal?	_____yes _____no
3. persuade the reader to examine this plan closely?	_____yes _____no
4. offer details about the content of the plan?	_____yes _____no
5. appear in full block form and include all six elements (return address, date, recipient's address, salutation, body, closing)?	_____yes _____no

Is the cover letter . . .

1. signed?	_____yes _____no
2. free of all grammatical and typographical errors?	_____yes _____no

Does the title page . . .

1. include all five elements (project title, name of sender name of recipient, date, return information)?	_____yes _____no
2. catch the attention of the reader?	_____yes _____no
3. have a title appropriate to the plan?	_____yes _____no

Is the title page . . .

1. visually appealing?	_____yes _____no
2. free of all grammatical and typographical errors?	_____yes _____no

What parts of the drafts did you like the most?

What parts of the drafts need the most improvement?

Abstract

1. Is the document clearly labeled as an "Abstract" at the top of the page? _____ yes _____ no

2. Is the document written from a third-person perspective? _____ yes _____ no

3. Does the document provide a balanced view of the project idea? _____ yes _____ no

4. Does the document cover elements of the problem, paradigm, and plan (in that order)? _____ yes _____ no

5. Does the document indicate a specific course of action? _____ yes _____ no

6. Is the document between 150 and 300 words and no longer than two paragraphs in length? _____ yes _____ no

7. Is the document single-spaced, in 12 point Times New Roman type? _____ yes _____ no

8. Is the document free of errors in grammar, usage, and/or sentence structure? _____ yes _____ no

9. Is the document presented in a clear, readable form? _____ yes _____ no

10. Would this document encourage me to read this plan? _____ yes _____ no

What is the one part of the draft you liked the most?

What is the one part of the draft that needs the most improvement?

Table of Contents and Table of Figures

1. Are these documents clearly labeled and presented in a logical and readable form? _____ yes _____ no

2. Are these documents free of errors in grammar, spacing, and punctuation? _____ yes _____ no

Additional Comments/Suggestions:

Chapter 7 ■ Final Proposal Workshop II

Please fill out the following form for your partner. Feel free to write comments on the drafts as well.

Introduction and Literature Review

Does the introduction . . .

1. attempt to quantify or define the problem? _____yes _____no

2. include visuals that help clarify and emphasize the key aspects of the problem? _____yes _____no

3. focus on the aspects of the problem that would most interest the reader? _____yes _____no

4. suggest a direction for approaching the problem? _____yes _____no

5. close with a forecasting statement giving the reader a sense of the argument to follow and providing a transition to the next section? _____yes _____no

Is the introduction . . .

1. single-spaced, in 12 point Times New Roman font? _____yes _____no

2. free of all grammatical and typographical errors? _____yes _____no

Does the literature review . . .

1. open with a reference to the problem? _____yes _____no

2. focus on the paradigm of the project? _____yes _____no

3. explain why the problem will be approached in a particular way? _____yes _____no

4. provide a unified rationale for the specific plan of action? _____yes _____no

5. show how the plan of action proposed is a logical approach to the problem defined? _____yes _____no

6. include the most useful and authoritative sources (especially those subject to peer review)? _____yes _____no

Is the literature review . . .

1. single-spaced, in 12 point Times New Roman font? _____yes _____no

2. free of all grammatical and typographical errors? _____yes _____no

What parts of the drafts did you like the most?

What parts of the drafts need the most improvement?

Additional Comments/Suggestions:

▓ Final Proposal Workshop III

Please fill out the following form for your partner. Feel free to write comments on the drafts as well.

Plan, Budget, and Discussion

Does the plan . . .

   1. transition logically from the research?   _____yes _____no

   2. focus on how the idea will be implemented?   _____yes _____no

   3. reference research to support the writer's choices?   _____yes _____no

   4. present information clearly?   _____yes _____no

   5. consider all possibilities in justifying its recommendations?   _____yes _____no

Is the plan . . .

   1. organized logically?   _____yes _____no

   2. free of unanswered questions or areas of confusion?   _____yes _____no

   3. single-spaced, in 12 point Times New Roman font?   _____yes _____no

   4. free of all grammatical and typographical errors?   _____yes _____no

Does the budget . . .

   1. list everything needed for the project?   _____yes _____no

   2. explain items that may be unfamiliar to the reader?   _____yes _____no

Is the budget . . .

   1. arranged in aligned accountant's columns?   _____yes _____no

   2. single-spaced, in 12 point Times New Roman font?   _____yes _____no

   3. free of all mathematical, grammatical, and typographical errors?   _____yes _____no

Does the discussion . . .

   1. conclude by summing up the project?   _____yes _____no

   2. make a final pitch for the value of the project?   _____yes _____no

   3. offer an evaluation plan for testing the results?   _____yes _____no

Is the discussion . . .

   1. single-spaced, in 12 point Times New Roman font?   _____yes _____no

   2. free of all grammatical and typographical errors?   _____yes _____no

Which parts of the drafts did you like the most?

Which parts of the drafts need the most improvement?

References

1. Is the document clearly labeled as a list of References at the top of the page? _____ yes _____ no

2. Does the document contain a minimum of ten published sources? _____ yes _____ no

3. Are there various types of sources represented (books to develop a theoretical framework, scholarly journals for detailed models, etc.)? _____ yes _____ no

4. Are at least 50% of the references cited as print sources? _____ yes _____ no

5. Is the document formatted in proper APA citation style (alphabetized, indented after first line, publication elements ordered correctly, etc.)? _____ yes _____ no

6. Is the document correctly spaced, in 12 point Times New Roman type, with one-inch margins? _____ yes _____ no

7. Is the document free of errors in grammar, punctuation, and capitalization? _____ yes _____ no

Additional Comments/Suggestions:

▪ Final Proposal Evaluation

1. The proposal includes all necessary sections and is within the page-length requirement.

   1   2   3   4   5   6   7   8   9   10

2. The proposal strives to persuade (and address the needs of) its audience.

   1   2   3   4   5   6   7   8   9   10

3. The proposal clearly describes and/or quantifies a viable problem, using published research and fieldwork.

   1   2   3   4   5   6   7   8   9   10

4. The proposal attempts a challenging and/or original task.

   1   2   3   4   5   6   7   8   9   10

5. The proposal is based upon relevant and/or innovative research.

   1   2   3   4   5   6   7   8   9   10

6. The References page includes the required number of sources and is presented in APA format.

   1   2   3   4   5   6   7   8   9   10

7. The research is organized into a clearly and carefully delineated paradigm.

   1   2   3   4   5   6   7   8   9   10

8. The plan of action follows logically from the research and is specifically described to the audience.

   1   2   3   4   5   6   7   8   9   10

9. The proposal places sources in logical relation to each other and to the project as a whole.

   1   2   3   4   5   6   7   8   9   10

10. The proposal is fully justified by the published research.

   1   2   3   4   5   6   7   8   9   10

11. The proposal engages possible complications suggested by the research or the plan.

   1   2   3   4   5   6   7   8   9   10

12. The transitions and headings help guide the reader through the project.

    1   2   3   4   5   6   7   8   9   10

13. The visuals are appropriate and effective at conveying information to the reader.

    1   2   3   4   5   6   7   8   9   10

14. The writing is fluent and virtually error-free.

    1   2   3   4   5   6   7   8   9   10

15. The proposal exhibits an overall attractive appearance and visually appealing design.

    1   2   3   4   5   6   7   8   9   10